Lecture Notes in Computer Science 14295

Founding Editors

Gerhard Goos
Juris Hartmanis

Editorial Board Members

Elisa Bertino, *Purdue University, West Lafayette, IN, USA*
Wen Gao, *Peking University, Beijing, China*
Bernhard Steffen, *TU Dortmund University, Dortmund, Germany*
Moti Yung, *Columbia University, New York, NY, USA*

The series Lecture Notes in Computer Science (LNCS), including its subseries Lecture Notes in Artificial Intelligence (LNAI) and Lecture Notes in Bioinformatics (LNBI), has established itself as a medium for the publication of new developments in computer science and information technology research, teaching, and education.

LNCS enjoys close cooperation with the computer science R & D community, the series counts many renowned academics among its volume editors and paper authors, and collaborates with prestigious societies. Its mission is to serve this international community by providing an invaluable service, mainly focused on the publication of conference and workshop proceedings and postproceedings. LNCS commenced publication in 1973.

Sharib Ali · Fons van der Sommen ·
Maureen van Eijnatten · Bartłomiej W. Papież ·
Yueming Jin · Iris Kolenbrander
Editors

Cancer Prevention Through Early Detection

Second International Workshop, CaPTion 2023
Held in Conjunction with MICCAI 2023
Vancouver, BC, Canada, October 12, 2023
Proceedings

Editors
Sharib Ali
University of Leeds
Leeds, UK

Maureen van Eijnatten
Eindhoven University of Technology
Eindhoven, The Netherlands

Yueming Jin
National University of Singapore
Singapore, Singapore

Fons van der Sommen
Eindhoven University of Technology
Eindhoven, The Netherlands

Bartłomiej W. Papież
University of Oxford
Oxford, UK

Iris Kolenbrander
Eindhoven University of Technology
Eindhoven, The Netherlands

ISSN 0302-9743 ISSN 1611-3349 (electronic)
Lecture Notes in Computer Science
ISBN 978-3-031-45349-6 ISBN 978-3-031-45350-2 (eBook)
https://doi.org/10.1007/978-3-031-45350-2

© The Editor(s) (if applicable) and The Author(s), under exclusive license
to Springer Nature Switzerland AG 2023

This work is subject to copyright. All rights are reserved by the Publisher, whether the whole or part of the material is concerned, specifically the rights of translation, reprinting, reuse of illustrations, recitation, broadcasting, reproduction on microfilms or in any other physical way, and transmission or information storage and retrieval, electronic adaptation, computer software, or by similar or dissimilar methodology now known or hereafter developed.
The use of general descriptive names, registered names, trademarks, service marks, etc. in this publication does not imply, even in the absence of a specific statement, that such names are exempt from the relevant protective laws and regulations and therefore free for general use.
The publisher, the authors, and the editors are safe to assume that the advice and information in this book are believed to be true and accurate at the date of publication. Neither the publisher nor the authors or the editors give a warranty, expressed or implied, with respect to the material contained herein or for any errors or omissions that may have been made. The publisher remains neutral with regard to jurisdictional claims in published maps and institutional affiliations.

This Springer imprint is published by the registered company Springer Nature Switzerland AG
The registered company address is: Gewerbestrasse 11, 6330 Cham, Switzerland

Paper in this product is recyclable.

Preface

CaPTion 2023 was the 2nd International Workshop on Cancer Prevention through early detecTion, organized as a satellite event of the 26th International Conference on Medical Image Computing and Computer Assisted Intervention (MICCAI 2023) in Vancouver, Canada. The main idea of founding CaPTion was to create a new research interface where medical image analysis, machine learning, and clinical researchers could interact and address the challenges related to cancer and early cancer detection using computational methods.

Early cancer diagnosis and its treatment for the long-term survival of cancer patients have been a battle for decades. 19.3 million cancer cases and almost 10 million deaths were reported in 2020, with lung (18%), colorectal (9.4%), liver (8.3%), stomach (7.7%) and female breast cancer (6.9%) being the leading causes of mortality. While computational methods in medical imaging have enabled the detection and assessment of cancerous tumors and assist in their treatment, early detection of cancer precursors opens an opportunity for early treatment and prevention. The workshop provided an opportunity to present research work in medical imaging around the central theme of early cancer detection. It strove to address the challenges that must be overcome to translate computational methods to clinical practice through well-designed, generalizable, interpretable, and clinically transferable methods. Through this new workshop, we aimed to identify a new ecosystem that would enable comprehensive method validation and reliability of methods, setting up a new gold standard for sample size and elaborating evaluation strategies to identify failure modes of methods when applied to real-world clinical environments.

The CaPTion 2023 proceedings contain 11 high-quality papers of between 10 and 14 pages selected through a rigorous peer review process (with an average of three reviews per paper). All submissions were peer-reviewed through a double-blind process by at least three members of the scientific review committee, comprising 21 experts (including chairs) in the field of medical imaging, especially of early cancer detection.

The accepted manuscripts cover various medical image analysis methods primarily focused on cancer and early cancer detection, progression, inflammation understanding, multimodality data, and computer-aided navigation. In addition to the papers presented in this LNCS volume, the workshop included three keynote presentations from world-renowned experts: Sir Mike Brady (University of Oxford, UK), Anne Martel (University of Toronto, Canada), and Sravanthi Parasa (Swedish Medical Center, Seattle, USA).

We wish to thank all the CaPTion 2023 authors for their participation and the members of the scientific review committee for their feedback and commitment to the workshop. We are very grateful to our sponsors Satisfai Health Inc. for their valuable support.

October 2023

Sharib Ali
Fons van der Sommen
Maureen van Eijnatten
Bartłomiej W. Papież
Yueming Jin
Iris Kolenbrander
Pedro Chavarrias

Organization

Program Committee Chairs

Sharib Ali — University of Leeds, UK
Fons van der Sommen — Eindhoven University of Technology, The Netherlands
Maureen van Eijnatten — Eindhoven University of Technology, The Netherlands
Bartłomiej W. Papież — University of Oxford, UK
Yueming Jin — University College London, UK
Iris Kolenbrander — Eindhoven University of Technology, The Netherlands

Student Representative Board Members

Iris Kolenbrander — Eindhoven University of Technology, The Netherlands
Pedro Chavarrias — University of Leeds, UK

Keynote Speakers

Sir Michael Brady — University of Oxford, UK
Anne Martel — University of Toronto, Canada
Sravanthi Parasa — Swedish Medical Center, Seattle, USA

Scientific Review Committee

Pedro Chavarrias — University of Leeds, UK
Christian Daul — Université de Lorraine, France
Mariia Dmitrieva — Queen Square Analytics, UK
Noha Ghatwary — University of Lincoln, UK
Mark Janse — UMC Utrecht, The Netherlands
Tim Jaspers — Eindhoven University of Technology, The Netherlands

Carolus Kusters	Eindhoven University of Technology, The Netherlands
Daneil Lang	Helmholtz Center Munich, Germany
Van Linh Le	University of Bordeaux, France
Gilberto Ochoa-Ruiz	Tecnológico de Monterrey, Mexico
Mansoor Ali Teevno	Tecnológico de Monterrey, Mexico
Christiaan Viviers	Eindhoven University of Technology, The Netherlands
Ziang Xu	University of Oxford, UK
Jiadong Zhang	ShanghaiTech University, China

Contents

Classification

Detection and Segmentation

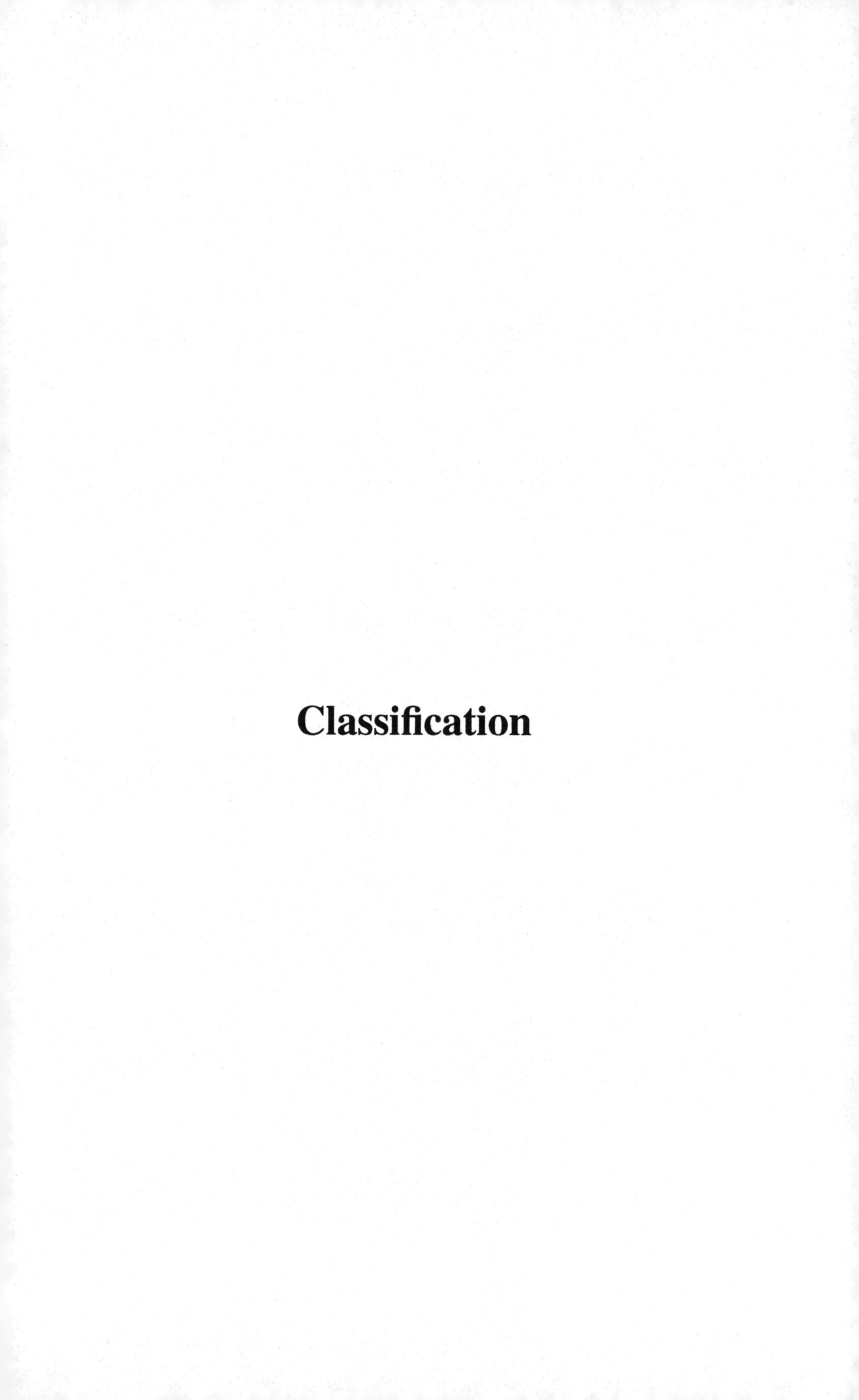

Classification

A Deep Attention-Multiple Instance Learning Framework to Predict Survival of Soft-Tissue Sarcoma from Whole Slide Images

Van-Linh Le[1,2,3,8](✉), Audrey Michot[3,5], Amandine Crombé[1,4,6], Carine Ngo[9], Charles Honoré[10], Jean-Michel Coindre[4,7], Olivier Saut[1,2], and Francois Le-Loarer[3,4,7]

[1] MONC Team, INRIA Bordeaux Sud-Ouest, Talence, France
[2] Bordeaux Mathematics Institute UMR 5251 (IMB), University of Bordeaux, CNRS and Bordeaux INP, Talence, France
van-linh.le@u-bordeaux.fr
[3] Bordeaux Institute of Oncology, BRIC U1312, INSERM, University of Bordeaux, Institute Bergonié, 33000 Bordeaux, France
[4] Faculty of Medicine, University of Bordeaux, Bordeaux, France
[5] Department of Oncological Surgery, Institute Bergonié, Bordeaux, France
[6] Department of Radiology, Pellegrin University Hospital, Bordeaux, France
[7] Department of Pathology, Institute Bergonié, Bordeaux, France
[8] Department of Data and Digital Health, Institute Bergonié, Bordeaux, France
[9] Department of Pathology, Institute Gustave Roussy, Villejuif, France
[10] Department of Oncological Surgery, Institute Gustave Roussy, Villejuif, France

Abstract. Soft-tissue sarcomas are heterogeneous cancers of the mesenchymal lineage that can develop anywhere in the body. A precise prediction of sarcomas patients' prognosis is critical for clinicians to define an adequate treatment plan. In this paper, we proposed an end-to-end Deep learning framework via Multiple Instance Learning (MIL), Deep Attention-MIL framework, for the survival predictions: Overall survival (OS), Metastasis-free survival (MFS), and Local-recurrence free survival (LRFS) of sarcomas patients, by studying the features from Whole Slide Images (WSIs) of their tumors. The Deep Attention-MIL framework consists of three steps: tiles selection from the WSIs to choose the relevant tiles for the study; tiles feature extraction by using a pre-trained deep learning model; and a Deep Attention-MIL model to predict the risk score for each patient via MIL approach. The risk scores outputted from the Deep Attention-MIL model are used to divide the patients into low/high-risk groups and predict survival time. The framework was trained and validated on a local dataset including 220 patients, then it was used to predict the survival for 48 patients in an external validation dataset. The experiments showed the proposed framework yielded satisfactory and promising results and contributed to accurate cancer survival predictions on both the validation and external testing datasets: By using the WSIs

O. Saut and F. Le-Loarer—Co-directed the research.

© The Author(s), under exclusive license to Springer Nature Switzerland AG 2023
S. Ali et al. (Eds.): CaPTion 2023, LNCS 14295, pp. 3–16, 2023.
https://doi.org/10.1007/978-3-031-45350-2_1

feature only, we obtained an average C-index (of 5-fold cross-validation) of 0.6901, 0.7179, and 0.6211 for OS, MFS, and LRFS tasks on the validation dataset, respectively. On the external testing dataset, these scores are 0.6294, 0.682, and 0.76 for the three tasks (OS, MFS, LRFS), respectively. By adding the clinical features, these scores have been improved both on validation and external testing datasets. We obtained an average C-index of 0.7835/0.6378, 0.7389/0.6885, and 0.6883/0.7272 for the three tasks (OS, MFS, LRFS) on validation/external testing datasets.

Keywords: Multiple Instance Learning · Deep Attention model · Survival prediction · Soft-tissue sarcoma · Whole Slide Image

1 Introduction

Soft-tissue sarcomas (STS) are heterogeneous malignant tumors developing anywhere in the body. They represent 1% of cancers in adults and 5% in children. STSs have a variable prognosis, their management requires the use of aggressive treatments including debilitating surgeries and/or high-dose chemotherapies. The prognosis of STS is dominated by two events: local recurrence and distant metastasis. The occurrence of metastasis is a major adverse factor for overall survival (OS), but local control of the disease also impacts OS [1]. In most studies, the most significant factor to predict local recurrence is the quality of surgical margins [1], whereas metastasis and overall survival are mostly related to the FNCLCC histological grade [2] which remains to date the most widely used standard to predict survival of sarcoma patients. A clinical nomogram integrating the grade and clinical variables such as patient age and tumor size has improved the prognosis evaluation of sarcomas patients [3].

In addition to biological and clinical information, Whole Slide Images (WSIs) contain information relevant for analyzing the diagnosis and prognosis of cancer, e.g., Overall survival (OS), Metastasis-free survival (MFS), or Local-recurrence free survival (LRFS), as well as prediction of response to treatment. WSIs can indeed assess the tumor growth and morphology in detailed, high resolution. However, capturing cell detail makes the exported image potentially cumbersome to cope with, and analyzing WSIs challenging for several reasons: (1) WSIs may contain a billion difficult pixels to process computationally; (2) a patient could have several WSIs for study, with significant differences in texture and biological structures; (3) we receive an only label at the patient level but different WSIs for diagnosis.

Deep learning has become a current solution for image processing applications comprising pathological image analysis. However, the processing of WSIs is different from usual images due to the massive resolution of this kind of image. One possible solution to overcome this challenge is to consider a weakly supervised method via a Multiple Instance Learning (MIL) approach [4]. In MIL, we split a WSI into non-overlapping tiles (patches). Therefore, a WSI could be considered a bag of tiles. It is not mandatory to analyze all tiles, as some of them may not be relevant for diagnostic detection; therefore, a subset of tiles is selected for the study.

In recent years, Deep Learning via MIL [5] has emerged as a promising way to predict survival in cancer patients by analyzing the WSIs [6,7]. Ilse et al. [6] have proposed to use the attention-MIL for classifying the histopathological images of breast and colon cancers datasets: the model with attention operator outperformed the other operators (e.g., max-pooling MIL or mean-pooling MIL), achieving an average AUC (of 5 folds cross-validation) of 0.775 and 0.968 in breast and colon cancers datasets, respectively. Yao et al. [7] have introduced a combination of a Siamese model [8] and an attention-based one to predict survival based on imaging features. The method used the K-Means algorithm [9] to cluster the imaging features into several phenotypes. Then, the Siamese model was used to extract the features for each phenotype before feeding to the attention module for aggregating the WSI-level feature. Finally, the aggregated WSI feature was processed by two fully-connected layers and outputted the risk score for each patient. The method was applied to lung and colorectal cancer datasets. The C-indexes on lung and colorectal datasets were 0.6963 and 0.652, respectively. Likewise, Pierre et al. [10] have proposed the MesoNet model to predict the OS of mesothelioma patients. To develop their model, they firstly splitted the WSIs into tiles with a size of 224 $\times$ 224 pixels and selected 10K of tiles for analysis due to the limitation of the computation memory. Secondly, the pre-trained ResNet50 [11] was used to extract the features of the tiles. Then, a convolutional one-dimensional was used to generate the score for each tile. Finally, the 10 highest and 10 lowest scores were selected and used as the input for the multi-layer perceptron classifier to provide the scores for each patient. MesoNet has achieved an average C-index of 0.642 and 0.643 on the training (2981 patients) and testing dataset (56 patients), respectively [10].

In this work, we report an end-to-end framework, Deep Attention-MIL, to predict the survival of sarcomas patients. At the heart of our framework is a deep learning model with an attention mechanism for survival prediction via MIL. We evaluate the proposed framework on two datasets originating from two different comprehensive cancer centers in France. In this work, we show that the framework offers satisfactory predictions of the survival probability of the patients compared to the gold standard used sarcomas patients, the FNCLCC histological grade [12] (Sect. 3.2).

2 Methodology

Figure 1 presents the workflows of the proposed framework. It was developed in three phases: firstly, the non-overlapping patches (tiles) were extracted from the WSIs of the patients. Then, a pre-trained deep learning model (e.g., ResNet50 [11]) was used as an encoder to extract the features from the tiles. Finally, the extracted features were fed into the deep learning survival model to predict the risk score for each patient. The risk scores were then taken by a non-parametric estimator (e.g., Kaplan-Meier [13]) to predict the survival probability for the patient.

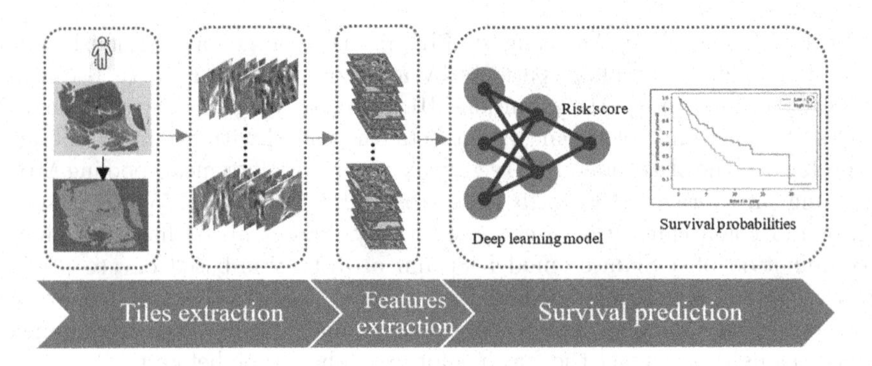

Fig. 1. The workflow of the proposed framework

2.1 Tiles Extraction

A WSI is organized as a pyramid of images with different magnification levels, the image at the highest level (40× of magnification) could have a resolution of $100K \times 100K$ pixels, and it is down-sampled over the magnification levels. Usually, the image at the highest magnification level (40×) is chosen for study. It is worth noting that a large part of the image does not contain any tissue, and is therefore useless for analysis. These regions are discarded during the tiling process. Because of the problem with high resolution of the original image, it is impossible to apply the classical image processing techniques on whole image for pre-processing step. Instead, an image at a lower magnification level (e.g., 12×, scaled image) is used to perform the pre-processing operations (e.g., segmentation, binarization), to select the tissue area, and to determine and mark the location of the interesting tiles. At the end of this stage, the tiles on the original image corresponding to the selected tiles on the scaled images are extracted and used for the study.

As a preferred size from the deep learning models for image classification [11,14], the original image is divided into non-overlapping tiles with a size of 224 × 224 (W × H) pixels. Based on the fraction of tissue, tiles are classified into 4 groups: Group A consists of the tiles that are composed of more than 80% of tissue, group B contains the tiles which have more than 10% and less than 80% of tissue, group C includes the tiles which have more than 10% of tissue, and group D of that tiles that do not contain tissue. In the context of this work, the tiles composed of more than 10% (from group A and B) of tissue have been used for our analysis. As mentioned, a patient could have several WSIs, even if we only extract tissue tiles, we can still get hundreds of thousands of tiles for each patient.

2.2 Features Extraction

Unlike segmentation and detection tasks in WSIs analysis [15–17], our framework predicts patient-level outcome aggregated from tile-level information. As pointed out in [18], training patch-based CNNs for weakly supervised learning is very time-consuming (several weeks), we propose to use features from pre-trained models instead of using CNNs to learn features from scratch. Here, for instance, we use a pre-trained ResNet50 [11] on ImageNet [14] as an extractor to extract the features of the tiles. For each tile, 2048 features are considered. Then, the extracted features are concatenated to obtain the features for each patient. The extracted features of WSIs are presented as a matrix of ($N \times 2048$), where N is the total number of WSI tiles.

2.3 Deep Attention-Multiple Instance Learning Model for Survival Prediction

Figure 2 illustrates the proposed deep learning architecture (Deep Attention-MIL) for survival predictions. The layers of this model can be decomposed into three groups: the first group consists of the layers before the Attention module [6]. These are used to aggregate the features for each tile as well as compute the score for each tile; the second group is an Attention module [6] which outputs the attention score for each tile. These scores present the importance of each tile conducted to the final prediction. The attention score is combined with the corresponding tile features before passing it to the layers in the last group to estimate the risk score for each patient.

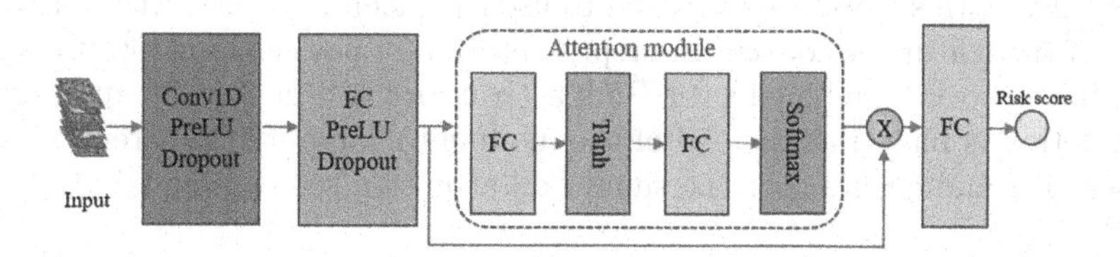

Fig. 2. The proposed Deep Attention-MIL survival model

Our model is derived from other studies [6,7] with modifications to adapt the architecture to our objective. First, we replaced the ReLU activation function with PreLU activation, which is more precise in the decision by making the leakage coefficient a parameter that is learned along with the other network parameters [19]. Then, to prevent overfitting, we added dropout layers [20] at the end of each group of layers, and reduce the number of features in the last layer of the model. It is worth that the final model was obtained after trying different combinations of layers and performing experiments on the same dataset.

Table 1 details the input/output dimensions at each layer/module of the proposed Deep Attention-MIL survival model. The input of the model is the selected features of the patient, organized as a matrix of ($M \times 2048$) where M is the number of tiles considered. After passing the layers in the first group, the features of each tile are collected and dimension is reduced to 256. These reduced features are inputted to the attention module to output the attention score for each tile

Table 1. The input/output dimensions at each layer of the survival model.

Layer	Input	Output
Conv1D/PreLU/Dropout	$M \times 2048$	$M \times 512$
FC/PreLU/Dropout	$M \times 512$	$M \times 256$
Attention module	$M \times 256$	1×256
FC	1×256	1

($1 \times M$). The attention score is multiplied with corresponding tile features to obtain the representation feature for WSI (1×256) which is the input for the last layer in the model. Finally, a linear function learns the representation of WSI to provide the risk score for the patient.

Attention Module: Local representation (two layers before the attention module) encodes features of the tiles, but our model provides the score at the patient level. Therefore, aggregating tile features into patient-level representation is a necessary step. A popular choice would be to use the maximum or the mean operator. Yet, the drawbacks are clear: they are pre-defined and not trainable which might not be adequate for this specific task. A better way to integrate tile information is to leverage an attention mechanism that considers the importance of each tile. In this work, we proposed to use the attention based MIL for aggregation of tile features to obtain the representation of patient-level [6]. It consists of two linear layers combined with Tanh activation functions. A soft-max activation function is placed at the end of the module to compute the attention score for each tile, and ensure that the sum of all attention scores is equal to 1.

3 Experiments and Results

3.1 Dataset Description and Experimental Setups

Datasets: The experiments were carried out on two different clinical cohorts from two comprehensive cancer centers: Institute Bergonié (Bordeaux, France) (IB dataset) and Gustave Rossy (Villejuif, France) (IGR dataset). These two cohorts were extracted from the Sarcoma BCB (https://sarcomabcb.org:connect). The criteria of inclusion included: primary sarcoma, location of trunk walls and limbs, upfront surgical resection, and patient naive of neoadjuvant therapy. The IB dataset consisted of 220 patients with more than 450 WSIs representing at least 2 WSIs per patient. The samples in this dataset were collected from 01/01/1990 to 01/12/2020. The IGR dataset consisted of 48 patients with more than 100 WSIs collected from 01/01/2000 to 01/12/2016. For all included patients, clinical follow-up was updated regarding survival, date of death, occurrence of metastasis and local recurrence. The IB dataset was used to train and validate the models. The IGR dataset was used as an external validation cohort (testing dataset). Table 2 details the number of patients, number of WSIs, recorded events of patients, location for collecting samples for each dataset.

Ethics: This study was conducted following local ethical guidelines and approved by the institutional research board of our institutions; all cases are recorded in the French expert sarcoma network (NetSarc+) database, which is approved by the National Committee for Protection of Personal Data (CNIL, no. 910390).

Table 2. The details of studied datasets.

Dataset	IB	IGR
No. patients	220	48
No. WSIs	450	105
No. patients alive/dead	133/87	36/12
No. patients non-metastatic/metastatic	148/72	37/11
No. patients non-recurrence/recurrence	188/32	40/8
Location	Institute Bergonié (Bordeaux, France)	Institute Gustave Roussy (Villejuif, France)

Experiments Setup: We evaluated the performance of the model on three survival tasks: Overall survival (OS), Metastasis free survival (MFS), or Local-recurrence free survival (LRFS). For each task, we performed a 5-fold cross-validation on the training dataset (IB). Then, the 5 corresponding models were used to predict the scores for the patients in the external validation set. (IGR). For all three tasks, we reported the C-index and Confidence Interval (CI-95%). The reported C-index in this study was the average C-index of 5-fold cross-validation.

As mentioned in Sect. 2.1, each patient had a hundred thousand tiles. For clear computational reasons, we could not analyze all tiles; therefore, we selected a subset of tiles (M = 10K) for the study. This value is empirically set after trying different values for the number of tiles for each patient. As the output of this step, each patient was represented as a matrix of ($M \times 2048$), this matrix was used as the input of the survival model.

Implementation Details: The model was implemented in the PyTorch library [21]. The model was trained for 200 epochs using an Adam optimization [22] with a weight decay of 10^{-4}. The learning rate and batch size have been set to 3×10^{-3} and 1, respectively. An early stopping strategy was applied by monitoring the validation loss to avoid over-fitting.

3.2 Experimental Results

In this section, we investigate the performance of our approach. First, we present the model's performances on tile features only for three survival tasks: OS, MFS, and LRFS. Then, we add more insight to the model by considering some clinical features, compare the two approaches (with and without clinical features). Finally, these results are compared to the results of a Cox model [23], a standard model for survival prediction.

Table 3. The C-index scores (± CI-95%) for OS, MFS, and LRFS tasks on IB validation and IGR testing datasets (with WSI features only).

Dataset	OS	MFS	LRFS
IB	0.6901 (± 0.0388)	0.7179 (± 0.0709)	0.6211(± 0.0537)
IGR	0.6294 (± 0.0153)	0.6820 (± 0.0491)	0.7600(± 0.0184)

Prediction from Tiles Features: Table 3 presents the obtained average C-index (±CI-95%) for each task on each dataset. On the validation set (IB), we obtained the average C-index scores of 0.6901 (±0.0388), 0.7179 (±0.0709), and 0.6211 (±0.0537) for OS, MFS, and LRFS tasks, respectively. On the external validation set (IGR), the average C-index scores on OS and MFS tasks are lower than the validation set, 0.6294 (±0.0153) and 0.682 (±0.0491) for OS and MFS, respectively. However, the C-index on LRFS outperforms the score on the validation set 0.76 (±0.0184). Although the C-index scores are a bit smaller on the test set, the difference is tiny. This could indicate a good generalization ability of our approach to unseen data.

As mentioned, the model provided the risk (event) score for each patient. Then, the risk scores were used to divide the patients into two groups: low-risk and high-risk, using Eq. 1. Finally, an estimator (e.g., Kaplan-Meier [13]) was used to obtain the survival probability of the two groups.

$$dx = f(x) = \begin{cases} 0 & \text{if } Px < PI \\ 1 & \text{if } Px \geq PI \end{cases} \tag{1}$$

where Px is the predicted risk score from the model and PI is the median of the risk scores.

Figure 3 illustrates the survival curves (in 10 years) of low/high-risk patients by using the Kaplan-Meier estimator [13] for each task on the IB validation dataset and the IGR testing dataset. On the IB validation dataset (left column in 3),the statistical information between the two groups was significant ($p < 0.05$), and the predictions of our model were good enough to separate the patients. The prediction curves were compared to the Grade curves (Grade is a gold standard to classify cancer tissues based on their appearance and behavior when viewed under a microscope for helping the doctor know about the aggressiveness of cancer. The grade is usually described using a number from 1 to 3 or 4. The higher the number, the more different the cancer tissues look from normal tissues and the faster they are growing), the curves provided by our scores are the same

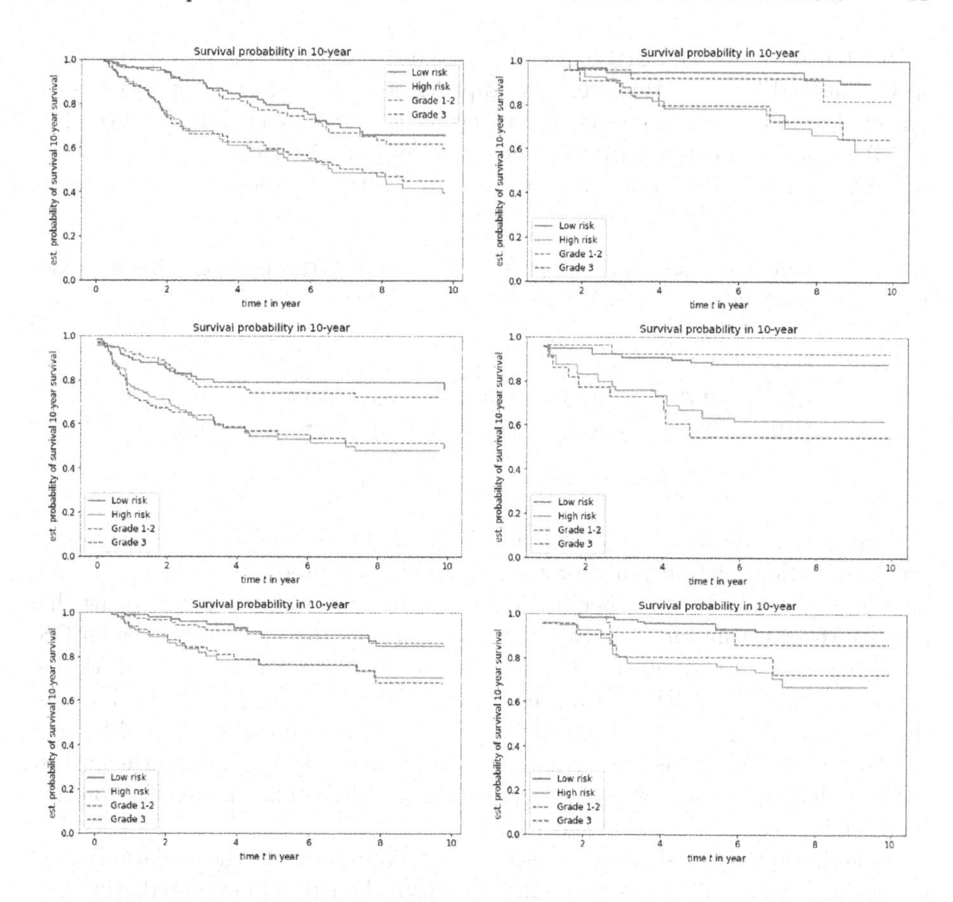

Fig. 3. The survival curves on IB validation (left column) and IGR testing (right column) datasets. From top to bottom: the survival curves on OS, MFS, and LRFS tasks compared to the histological Grade.

level or better in some periods (e.g., from 5 to 6 years). On the external cohort (IGR) (right column in Fig. 3), the stratification for OS and MFS were similar to grade and a bit better for LRFS.

Adding Clinical Features to Imaging: This section presents the results of the enhanced version of the proposed deep learning model. In this version, we consider additional clinical features beside the imaging features. In order to compare to another approach on clinical features, clinicians have selected four clinical features: *age, size of tumor, grade, and histotype*, to add to the tiles features for the prediction of survival.

Table 4 summaries the C-index ($\pm$ CI-95%) for each task on each dataset. On the validation dataset (IB), the C-index is improved on all three survival tasks: OS - 0.7835 ($\pm$0.034), MFS - 0.7389 ($\pm$0.034), LRFS - 0.6883 ($\pm$0.037). On IGR dataset, the scores for three survival tasks are: OS - 0.6378 ($\pm$0.043),

MFS - 0.6885 (±0.033), LRFS - 0.7272 (±0.068). It is the same improvement on OS and MFS tasks. However, the improvement in IGR was not as large as expected from the validation set results, even the score was decreased a little bit on LRFS task. To explain the difference, we hypothesized that we have a bias between the clinical data of the patients from the two centers.

Table 4. The C-index scores (± CI-95%) of OS, MFS, LRFS tasks on IB validation and IGR testing datasets (with WSI and clinical features).

Dataset	OS	MFS	LRFS
IB	0.7835 (± 0.034)	0.7389 (± 0.034)	0.6883 (± 0.037)
IGR	0.6378 (± 0.043)	0.6885 (± 0.033)	0.7272 (± 0.068)

Using the same strategy to evaluate the prediction scores as presented in the previous section, the output scores of the model are used to split the patients into low/high-risk groups. Then, we illustrate the two group curves by using the Kaplan-Meier estimator (Fig. 4). On the IB dataset, the survival curves on OS and MFS tasks are significantly separate, and they are the same level as the grade curves; on the LRFS task, the model met difficulty to split the patients in the first period of 5 years. The survival curves (for three tasks: OS, MFS, and LRFS) on the IGR dataset are not significantly changed compared to the curves without clinical features. It seems that adding the clinical has more effect on the IB (C-index scores) than the IGR dataset.

In addition to the risk score, proposed model also provides the attention score for each tile, which allows us to predict the survival pattern before predicting the risk score. Figure 5 illustrates the top 15 tiles of a patient who had a metastatic relapse, these tiles with high attention scores are the ones affecting the most the model prediction. Among these, 4 interested normal tissue surrounding the tumor, 9 originated from the tumor, and 2 normal tissue far from the tumor.

Comparison with the Cox Model: Cox model [23] is a popular model for the prediction of survival. We have used the Cox model with two objectives: (1) to verify the informative value of tile encoding, is it enough for use as a feature in a survival model? (2) to compare the performance of our framework with a classical method for survival tasks.

Table 5. The C-index scores from Cox model on OS, MFS, LRFS tasks.

Dataset	OS	MFS	LRFS	Nb. features
IB	0.7455	0.6835	0.7216	4 features (clinical)
IGR	0.57	0.7049	0.7381	
IB	0.7591	0.7071	0.7274	5 features (clinical + risk score)
IGR	0.5733	0.74	0.7279	

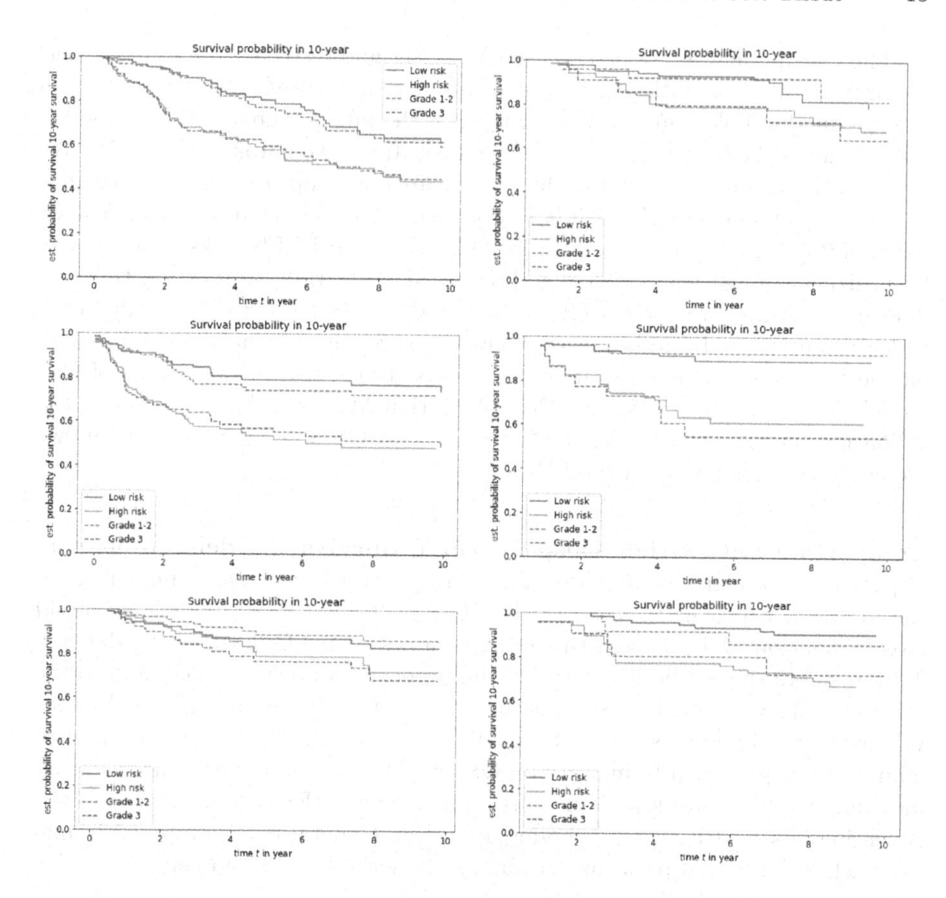

Fig. 4. The survival curves (in 10 years) on IB (left column) and IGR (right column) datasets by using imaging and clinical features. From top to bottom: the survival curves on OS, MFS, LRFS tasks.

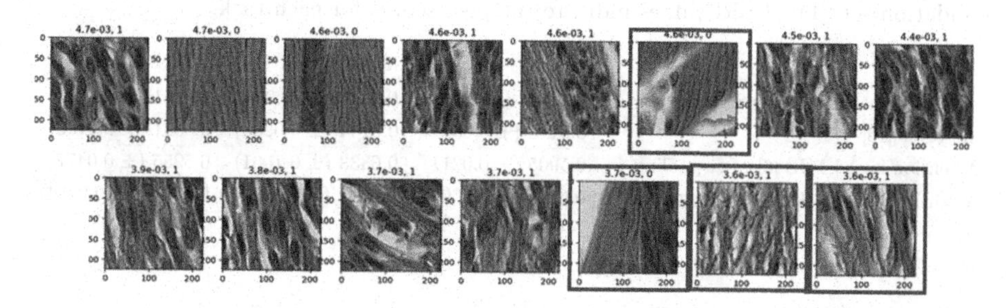

Fig. 5. Top-15 tiles with highest attention score of a metastasic patient from IB dataset. The blue boxes highlights tiles of normal tissue surrounding the tumor tissue. (Color figure online)

The Cox model was trained on the IB dataset and tested on the IGR dataset. We first try the Cox with 4 clinical features. Then, we used the outputs of deep learning model (risk score) on WSI features as the fifth feature on the Cox model. Table 5 shows the C-index scores on three tasks from two datasets (IB/IGR). We see from the results that adding the tile features has improved the performance of the Cox model. On the validation set, the Cox model has obtained a C-index of 0.7455, 0.6835, and 0.7216 for OS, MFS, and LRFS tasks, respectively, by utilizing clinical features. By adding the deep learning risk score, these C-indexes have improved to 0.7591, 0.7071, and 0.7274 for OS, MFS, and LRFS tasks, respectively. In addition, the C-index scores have been improved as well on the testing set. Besides that, in the comparison between the results of Cox model (5 features) and proposed Deep Attention-MIL model, our results on the validation set are better than the Cox model on OS and MFS tasks, while we are close to Cox's result on the LRFS task.

Comparison with Other Deep Learning Survival Models: To have an objective assessment of the performance of the model, we have re-implemented the methods which have described in [10] and [6]. Then, we performed 5-fold cross-validation and reported the average values of the C-index on IB dataset. Table 6 shows the prediction power of the proposed framework compared to the different survival models based on the average C-index scores of 5-fold cross-validation on different survival tasks. We see from the table that the performance of our proposed framework on OS and MFS tasks outperforms the other methods, while we are less than a little bit on the LRFS task compared to the Attention-based MIL average pooling approach. Generally, the proposed framework achieves the best performance among all methods on most of survival tasks.

Table 6. Performance comparison of the proposed framework with other available methods using average C-index scores (± CI-95%) of OS, MFS, and LRS tasks on IB validation set. The **bold** values indicate the best scores for each task.

Method	OS	MFS	LRFS
Deep Attention-MIL (proposed)	**0.6901** (± 0.0388)	**0.7179** (± 0.0709)	0.6211 (± 0.0537)
MesoNet [10]	0.6118 (± 0.0531)	0.6134 (± 0.0283)	0.5881 (± 0.0790)
Attention-based MIL [6] Max-pooling	0.5905 (± 0.0211)	0.6333 (± 0.1061)	0.5255 (± 0.0477)
Attention-based MIL [6] Average-pooling	0.6468 (± 0.0609)	0.6535 (± 0.0671)	**0.6444** (± 0.0454)

4 Discussion and Conclusion

In this paper, we propose a Deep Attention-MIL framework for survival predictions from WSIs. Our objective was to investigate the role of the WSIs for the prognostic tasks along the clinical features. First, we have built a baseline from the Cox model, which took into account the clinical only. Then, the deep learning risk score obtained from WSIs features was added to the analysis of the Cox model. In this study, we show that imaging adds some fascinating insight into the Cox analysis, and that it can improve the performance of the Cox model. Finally, we propose a fully deep learning framework to combine the tile and clinical features for survival prediction in sarcoma patients. Our framework achieved good performance for various survival prediction tasks, even better results than the Cox model, and comparable to the gold standard for cancer studies. The results have been also compared to several deep survival models, the comparison showed that our framework achieves higher performance than these methods. One should keep in mind that this model may apply to sarcoma patients affected with any sarcoma histotypes developed in the trunk walls and the limbs naive of treatment.

The present study raises some questions that we plan to address in future works. (1) Concerning the feature extraction step, one may replace the current extractor with another extractor that retains a relation with studied images, for example, we are examining a self-supervised learning model which can be downstream to use as an extractor. (2) We plan on analyzing the effect of the origins of the tiles on the tile selection step, in which we are ongoing to automate the classification of the tiles from different regions. (3) One needs to investigate the variability of the WSIs coming from various centers to improve the model's performances and develop an adequate harmonization method.

Acknowledgements. This work was supported by grants from the SIRIC Bordeaux and the Fondation Bergonié. This study is carried out in collaboration between the MONC team, Inria Bordeaux Sud-Ouest, and the BRIC U1312.

References

1. Casali, P.G., et al.: Soft tissue and visceral sarcomas: Esmo-euracan clinical practice guidelines for diagnosis, treatment and follow-up. Ann. Oncol. **29**, iv51–iv67 (2018)
2. Coindre, J.-M., Terrier, P., Bui, N.B., et al.: Prognostic factors in adult patients with locally controlled soft tissue sarcoma: a study of 546 patients from the French federation of cancer centers sarcoma group. Journal of Clinical Oncology **14**(3), 869–877 (1996)
3. Callegaro, D., et al.: Development and external validation of two nomograms to predict overall survival and occurrence of distant metastases in adults after surgical resection of localised soft-tissue sarcomas of the extremities: a retrospective analysis. Lancet Oncol. **17**(5), 671–680 (2016)
4. Herrera, F., et al.: Multiple Instance Learning. Springer, Heidelberg (2016). https://doi.org/10.1007/978-3-319-47759-6
5. Dietterich, T.G., Lathrop, R.H., Lozano-Pérez, T.: Solving the multiple instance problem with axis-parallel rectangles. Artif. Intell. **89**(1), 31–71 (1997)
6. Ilse, M., Tomczak, J., Welling, W.: Attention-based deep multiple instance learning. In International Conference on Machine Learning, pp. 2127–2136. PMLR (2018)
7. Yao, J., Zhu, X., Jonnagaddala, J., Hawkins, N., Huang, J.: Whole slide images based cancer survival prediction using attention guided deep multiple instance learning networks. Med. Image Anal. **65**, 101789 (2020)

8. Chopra, S., Hadsell, R., LeCun, Y.: Learning a similarity metric discriminatively, with application to face verification. In: 2005 IEEE Computer Society Conference on Computer Vision and Pattern Recognition (CVPR 2005), vol. 1, pp. 539–546. IEEE (2005)
9. MacQueen, J., et al.: Some methods for classification and analysis of multivariate observations. In Proceedings of the Fifth Berkeley Symposium on Mathematical Statistics and Probability, Oakland, CA, USA, vol. 1, pp. 281–297 (1967)
10. Courtiol, P., et al.: Deep learning-based classification of mesothelioma improves prediction of patient outcome. Nat. Med. **25**(10), 1519–1525 (2019)
11. He, K., Zhang, X., Ren, S., Sun, J.: Deep residual learning for image recognition. In Proceedings of the IEEE Conference on Computer Vision and Pattern Recognition, pp. 770–778 (2016)
12. Coindre, J.M., et al.: Predictive value of grade for metastasis development in the main histologic types of adult soft tissue sarcomas: a study of 1240 patients from the French federation of cancer centers sarcoma group. Cancer: Interdisc. Int. J. Am. Cancer Soc. **91**(10), 1914–1926 (2001)
13. Kaplan, E.L., Meier, P.: Nonparametric estimation from incomplete observations. J. Am. Stat. Assoc. **53**, 457–481 (1958)
14. Russakovsky, O., et al.: ImageNet large scale visual recognition challenge. Int. J. Comput. Vision (IJCV) **115**(3), 211–252 (2015)
15. Courtiol, P., Tramel, E.W., Sanselme, M., Wainrib, G.: Classification and disease localization in histopathology using only global labels: a weakly-supervised approach. preprint arXiv:1802.02212 (2018)
16. Rony, J., Belharbi, S., Dolz, J., Ayed, I.B., McCaffrey, L., Granger, E.: Deep weakly-supervised learning methods for classification and localization in histology images: a survey. arXiv preprint arXiv:1909.03354 (2019)
17. Wang, D., Khosla, A., et al.: Deep learning for identifying metastatic breast cancer. arXiv preprint arXiv:1606.05718 (2016)
18. Hou, L., Samaras, D., Kurç, T.M., Gao, Y., Davis, J.E., Saltz, J.: Efficient multiple instance convolutional neural networks for gigapixel resolution image classification, vol. 7, pp. 174–182 (2015). preprint arXiv:1504.07947
19. He, K., Zhang, X., et al.: Delving deep into rectifiers: surpassing human-level performance on imagenet classification. In: Proceedings of the IEEE International Conference on Computer Vision, pp. 1026–1034 (2015)
20. Srivastava, N., Hinton, G., Krizhevsky, A., Sutskever, I., Salakhutdinov, R.: Dropout: a simple way to prevent neural networks from overfitting. J. Mach. Learn. Res. **15**(1), 1929–1958 (2014)
21. Paszke, A., et al.: Automatic differentiation in pytorch. In: NIPS-W (2017)
22. Kingma, D.P., Ba, J.: Adam: a method for stochastic optimization. arXiv:1412.6980 (2014)
23. Cox, D.R.: Regression models and life-tables. J. Royal Stat. Soc. Ser. B (Methodological) **34**(2):187–202 (1972)

Towards Real-Time Confirmation of Breast Cancer in the OR Using CNN-Based Raman Spectroscopy Classification

David Grajales[1,2(✉)], William Le[1,2], Frédérick Dallaire[1,2], Guillaume Sheehy[1,2], Sandryne David[1,2], Trang Tran[1,2], Frédéric Leblond[1,2,3], Cynthia Ménard[2], and Samuel Kadoury[1,2]

[1] Polytechnique Montréal, Montréal, QC, Canada
david-orlando.grajales-lopera@polymtl.ca
[2] Centre de recherche du Centre Hospitalier de l'Université de Montréal, Montréal, QC, Canada
[3] Institut du Cancer de Montréal, Montréal, QC, Canada

Abstract. Breast-conserving surgery is a recommended treatment for early-stage breast cancer. Recurrence and post-operative complications are potential risks when margins are not entirely removed during surgery or when timing constraints in the OR limit extensive analysis of resected tissue. Raman spectroscopy (RS), a non-destructive optical technique, enables the acquisition of molecular signatures of tissue samples allowing confirmation of different diseases, including cancer. Typically, the measured spectra must be processed and used to train conventional machine learning classifiers for cancer/normal discrimination. However, there is a lack of real-time spatially-resolved information that allows confirmation of cancer at a specific site during surgery. In this paper, we propose a tissue characterization pipeline based on convolutional neural networks (CNN), using 4 × 1D convolutional layers for automated feature extraction and a fully-connected layer as an alternative to classifying the complete RS spectra (without previous feature selection). Using 169 samples collected from 20 patients, we evaluated the performance of the proposed model, achieving an accuracy and sensitivity of 0.93(0.01) and 0.94(0.02), respectively, improving over traditional SVM-based models. Results demonstrate the potential of CNN models for classification in the OR and highlight the value of efficient signal processing to enhance their use for in-situ cancer detection.

Keywords: Raman spectroscopy · Breast cancer · Convolutional neural networks

1 Introduction

It is estimated that each year, 287,850 women are diagnosed with invasive breast cancer in the US alone [9]. Breast-conserving surgery (BCS) is one of the most

© The Author(s), under exclusive license to Springer Nature Switzerland AG 2023
S. Ali et al. (Eds.): CaPTion 2023, LNCS 14295, pp. 17–28, 2023.
https://doi.org/10.1007/978-3-031-45350-2_2

suitable surgery techniques for early-stage cancer for both ductal carcinoma in situ (DCIS) and invasive cases [11,14,22]. The success rate of standard of care procedures depends considerably on the presence of residual tumor in the remaining mammary gland. However, due to the difficulty in visual discrimination of healthy from invasive cancer cells, incomplete excision can lead to recurrence and require additional surgery. While macroscopic evaluation of the margins is often performed intraoperatively, positive margins are still detected in up to 20–35% of cases during postoperative staging [11,12,14,24].

Currently, the lack of accurate real-time intraoperative techniques to detect cancer cells at the margins has motivated several studies to evaluate various tissue properties [17,18,24]. One such technique, Raman spectroscopy (RS), characterizes microscopic information providing real-time molecular signatures by taking advantage of the tissue's highly sensitive optical imaging properties [2,8]. Based on inelastic light scattering, RS has proven useful for ex-vivo characterization of several diseases including Alzheimer's, cardiovascular diseases and brain cancer [1,5,16], despite being a weak signal [13,23]. Furthermore, the development of optical fiber RS probes and the imaging modality's non-destructive nature favor its clinical usage [1,25] by facilitating the integration into existing workflows [10]. Nevertheless, the major limitation of integrating RS for true real-time analysis within the operating room (OR) is its dependency on domain-specific signal processing and feature selection steps [23].

With the generated spectral signatures, classification based on the distinctive hand-selected peaks is typically done using machine learning methods such as logistic regression [11] and support vector machines (SVM) [7,13,16,27], such as in [4] for breast cancer detection, reporting sensitivity and specificity of 0.92 and 0.90 respectively. Due to the high number of features, dimensionality reduction based on principal component analysis [7,27] or feature selection using Lasso regression are commonly used [10,15,21]. In recent years, autoencoders (AE) have seen growing use in several medical image processing applications [13,22]. Nonetheless, these current methods lack the task-specific automatic feature extraction properties that state-of-the-art methods such as deep learning (DL) provide for high-dimensional inputs.

The challenges to achieve real-time breast cancer detection are two-fold: 1) improving feature selection to maximize cancer cell discrimination and 2) reducing the run-time requirements to allow real-time RS analysis. Currently, both challenges require significant human involvement in traditional workflows. We propose a novel method for invasive breast cancer cell detection using an RS system for rapid tissue characterization by leveraging a 1D CNN for automatic feature extraction and classification. We compare this method against 2 SVM-based approaches on a dataset of 20 patients. Furthermore, we hypothesize that raw RS signals can adequately be used as inputs without significant loss in test sensitivity thus achieving the necessary requirements for its inclusion in the OR for breast cancer margin confirmation.

2 Related Work

In 2020, Santilli et al. [22] explored a DL approach for basal cell cancer detection using RS. In their study, the authors proposed an autoencoder (5 -fully connected layer encoder and 5 -fully connected layer decoder) to reconstruct the input signal and extract a latent space for classification. Working with 127 normal and 63 cancerous tissue samples, they applied signal processing and data augmentation to obtain 4000 samples in total, with which they trained a model obtaining an accuracy of 0.96 in binary classification. Moving to breast cancer, in 2021, Ma et al. [17] proposed a binary classification model using a single 1-dimensional CNN layer and two dense layers; in a cohort of 20 patients, they applied signal processing and data augmentation before the model training to obtain 0.92 in accuracy. In their study, Fisher Discrimination Analysis (FDA) and SVM (using different kernels) classifiers were trained and tested on the same data for comparison.

In other fields, such as mineral and bacterial classification, where data are not as limited, Zhou et al. [28] trained a deeper and more complex CNN-based model for multiclass discrimination (top-1 accuracy of 0.92). Inspired by the ResNet architecture, the authors propose a block consisting of 4 CNN layers, dense layers, activation functions, and an identity shortcut, a block used 3 times, along with other CNN and dense layers, to form the complete model. The mineral dataset contained more than 5000 samples, and the bacteria dataset 60 000; thus, data augmentation was not applied. The authors applied a relatively simpler signal processing since human body tissue's significant autofluorescence was absent, and emphasized the challenge and potential advantages of working with raw signals rather than processed data.

3 Materials and Methods

3.1 Clinical Data and Setup

The cohort in this study consisted of 20 breast cancer patients: 19 with a confirmed diagnosis of invasive cancer who underwent open breast surgery (mastectomy or lumpectomy) and one undergoing breast reduction surgery. For the latter, optical measurements were acquired on the breast in which no tumor was radiologically detectable while still diagnosed with breast cancer associated with a tumor, but detected in the contralateral breast. Informed consent was obtained before the patient underwent surgery (McGill University Health Center Ethics Committees, approval number 2021–5997). All recruited patients had a cancer grade inferior to 4 and a tumor larger than 1 cm. The RS system used for the ex-vivo optical measurements (Reveal Surgical, Montreal, Canada), consists of a light source (785 nm laser), a high sensitivity spectrometer (wavelengths from 800 nm to 900 nm) and an optical probe connected to the previous two components. This probe was a hand-held single-point RS probe system, which integrates ten optical fibers with a 100 μm diameter core: one central fiber used to stimulate the tissue and nine used to collect tissue response. This setup allowed the mesoscopic in-contact characterization of circular shape sites of 0.5 mm in diameter. More details on the system can be found in previous works [4,6].

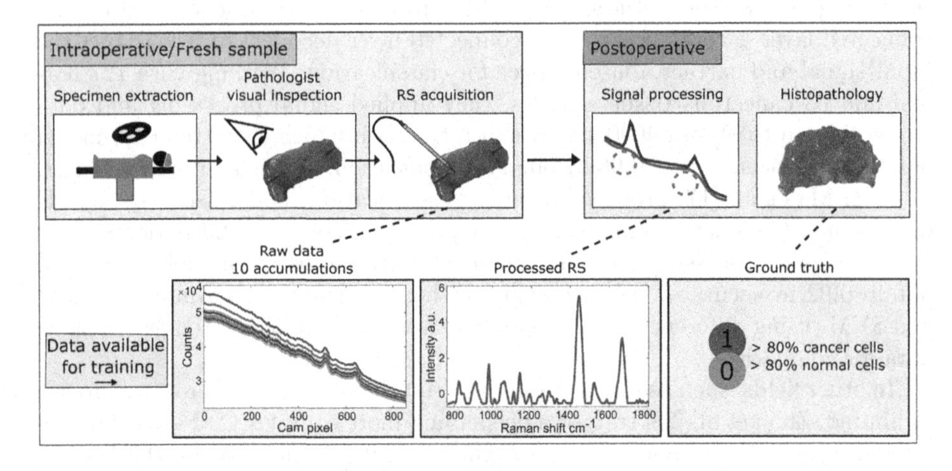

Fig. 1. Stages of the data acquisition process. The process goes from BCS to histopathological analysis and includes handling of fresh samples and signal processing. The bottom row presents the data used for training the RS classification model for cancer/normal discrimination, including an example of the 10 raw accumulations and the processed spectrum for one inspected site.

3.2 Workflow and Data Acquisition

For training purposes, one fresh specimen per patient was inked and sliced according to institution standards to obtain 5 mm thick slices. A pathologist then selected 2 smaller samples (cancer and normal) per slice based on visual inspection.

Several optical measurements were taken on different sites of each sample. A measurement consisted of 10 accumulations (repeated measurements) acquired at each location by setting the laser power to 100 mW with an exposure time per spectrum ranging from 0.1 to 4 s; a background measurement was also made before any accumulations with the laser off. More details on the selection of these parameters and the interaction with the tissue can be found at [4,6].

After RS, samples were fixed and processed according to standard histopathological procedures and observed by an expert to identify cells on the stained sample slides. The cells were reported as cancer (tumor cells, tumor stroma, or necrosis), normal (connective tissue, stroma, fibroblast, collagen), or fat (adipose cells) on every slide. For this study, the "Normal" label was attributed to the inspected site when at least 80% of the inspected area contained normal cells, and "Cancer" when 80% of the surface or more were cancer cells. Samples failing to meet these criteria, as well as fat samples, were excluded from the study.

The 10 accumulations were processed to obtain a single Raman spectrum per site (see Fig. 1). This process includes averaging the accumulations, background and cosmic rays subtraction, normalization with a NIST Raman standard (SRM 2214), autofluorescence removal, standard normal variate normalization, and assigning each pixel of the spectrometer to a Raman shift. A quantitative quality factor (QF) metric (range: 0–1) was computed from each resulting spectrum. Spectra with a QF metric inferior to 0.6 were excluded [3,23].

3.3 Classification Model

In contrast to the widely used SVM methods for binary classification of RS for cancer detection [8,20,21,27], DL approaches have only begun to be explored in recent years [13,17,28]. We propose a CNN-based model, fed with complete RS signals, to be compared with some SVM-based approaches.

Proposed Model: 1D CNN. In the proposed end-to-end DL-based approach, the selected architecture consisted of 4 × 1D convolutional layers, each followed by a batch normalization and ReLU activation function (see Fig. 2B) for the automated feature extraction. Specifically, 1D convolutions are well suited to detect salient peak signals and automatically extract/generate features from complete signals [22,28]. The number of layers (3) and features (120, 60, 30) in each were selected to maximize classification performance while minimizing model capacity to better adapt to smaller datasets inherent to a pilot study. For the discrimination component of the CNN, a fully-connected layer (30 features) allowed direct training of the CNN for supervised binary classification. The paradigm taking raw RS signals as inputs is the fastest of all compared methods as the fully end-to-end CNN, after training, can perform inference in near real-time (<1 s). Note that the paradigm taking processed RS signals as inputs still requires manual intervention to correctly tune the preprocessing steps.

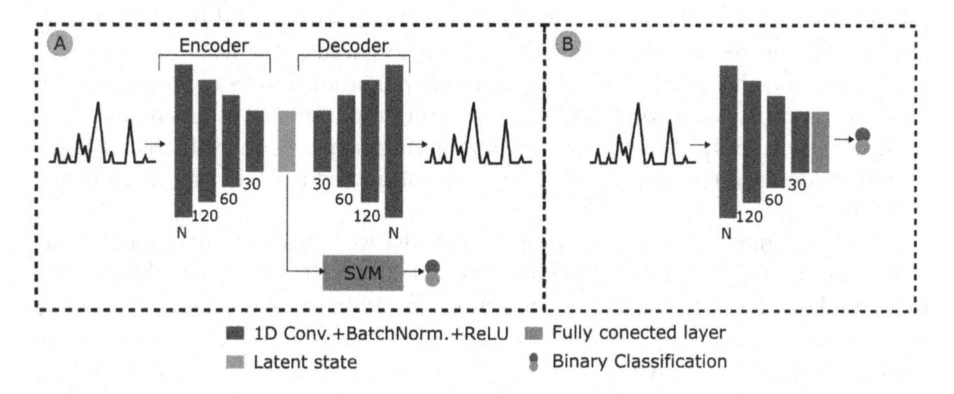

Fig. 2. Architecture of the DL-based models implemented for discriminating cancer from normal tissue using breast cancer RS: (A) alternative AE+SVM comparative method and (B) proposed 1D CNN. The baseline SVM model which is the same as in [4] is not shown.

Comparative Methods: SVM/AE+SVM. Previous studies have shown the classification power of 4 specific peaks on the processed Raman fingerprint (940 cm^{-1}, 1004 cm cm^{-1}, 1129 cm cm^{-1} and 1155 cm cm^{-1}) with a feature selection process based on Lasso regression [4]. We re-implemented the SVM methodology following the parameters used by *** et al. [4] to establish our baseline performance. An alternative method (AE+SVM) used an AE module for dimensionality reduction (see Fig. 2A). It was initially trained offline to craft features from the raw or processed RS signal. Here, both encoders and decoders consisted of 4 $\times$ 1D convolutional blocks as in the proposed fully CNN method with 120, 60, 30 and 30, 60 120 features respectively. In the online training phase, the decoder component was discarded and the encoder was fixed to train the SVM.

Implementation Details. In the proposed end-to-end DL approach, the 1D CNN was trained for 30 epochs with Adam optimizer and a learning rate of 0.001. All convolutional layers used a kernel size of 3 and a stride of 2 as a downsampling strategy. Models were trained on an NVIDIA-SMI GPU with 16GB RAM, optimizing for cross-entropy (1D CNN) loss function and implemented using PyTorch [19].

The AE component was trained initially offline for 50 epochs with the Adam optimizer and a learning rate of 0.001. Convolutional and transposed convolutional layers with kernel size 3 and stride 2 were used for downsampling and upsampling respectively. The minimization loss was the L1 loss for input signal reconstruction. In the online training phase, the 30-feature latent embedding was then used as inputs to the SVM model, implemented using Scikit-learn [26], and trained using a linear kernel in both SVM and AE+SVM methods. To reduce bias from training on an unbalanced dataset (see Table 1) the cost function penalized false positives at double the weight.

Training and Evaluation. A 5-fold patient cross-validation was performed for all experiments. The predictions and probabilities were used to compute the area under the receiver-operating-characteristic curve (AUC), accuracy, sensitivity, and specificity. The complete training-validation process was repeated 5 times for significance testing. Thus metrics were reported as the mean and standard deviation (SD) of 5 repetitions. The performances were evaluated using a paired Student's t-test; a p-value <0.05 was considered a statistically significant difference.

The hyperparameters of the proposed model were selected by a small ablation study in which different learning rates (0.001, 0.005, 0.01), epochs (20, 30, 60), number of layers (2, 4), and features ([n, 120, 60, 30], [n, 120, 120, 30]) were evaluated (see Appendix), with the results showing only the best model (comparison based on AUC).

4 Results

4.1 Clinical Data and Data Processing

In total, the dataset consisted of 169 inspected sites, with the corresponding histopathological report used as ground truth (for training purposes) and spectroscopy features (see Table 1). Figure 3A–B shows the raw and processed signals, averaged by the target class.

Table 1. Characteristics of the breast cancer patient cohort undergoing open breast surgery for invasive carcinoma. Each acquisition site (spectral measurement) consisted of 10 raw spectra, totaling 169 total sites.

Age at diagnosis (Cohort n = 20)	
Median (years)	67 (range: 54–77)
Number of sites acquisition	
Median (per patient)	6 (range: 1–25)
Site label distribution	
Normal (> 80% normal cells)	59
Cancer (> 80% cancer cells)	110
Number of spectral points	
Raw RS	854†
Processed RS	631

† Raw sample consists of 10 accumulations, and 854 features per accumulation.

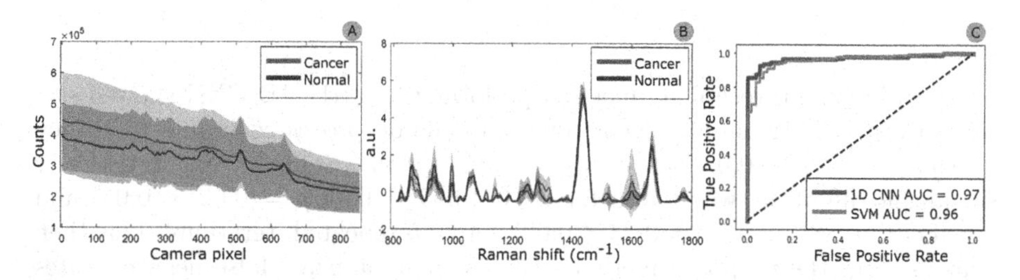

Fig. 3. End-to-end RS signal transformation to obtain breast cancer tissue classification. The (A) raw and (B) processed RS from 169 inspected sites were averaged by class and plotted with the solid line; the shaded area represents the SD. (C) The ROC curve for the proposed model (1D CNN) and the baseline (SVM).

4.2 Classification Performance

Table 2 presents the performance of the proposed 1D CNN method compared with the baseline (SVM) and SVM with dimensionality reduction (AE + SVM) methods. Furthermore, the columns present the results of the 3 methods trained on the standard processed RS inputs or the unprocessed signals (raw) aiming to minimize run-time delays. Note that current state-of-the-art breast cancer detection methods use SVM with processed signals and non-automatic feature selection, obtaining an accuracy of 0.91 during ex-vivo experiments [4] and >0.96 in more controlled in-vitro experiments [27].

Table 2. Comparative results of the proposed end-to-end DL model against other methods (5-fold, 5 repetitions cross-validation) for discriminating cancer from normal tissue. Bold font indicates top-performing results from the respective category (column). Asterisks indicate statistically significant improvement of the proposed model compared SVM baseline. All figures are presented as: mean(SD).

Model	Metric	Raw RS	Processed RS
SVM	AUC	0.82 (0.06)	0.95 (0.01)
	Accuracy	0.78 (0.03)	0.90 (0.01)
	Sensitivity	**0.88 (0.03)**	0.89 (0.01)
	Specificity	0.57 (0.07)	**0.91 (0.02)**
AE + SVM	AUC	0.77 (0.04)	0.89 (0.01)
	Accuracy	0.72 (0.02)	0.86 (0.02)
	Sensitivity	0.83 (0.03)	0.90 (0.03)
	Specificity	0.51 (0.16)	0.77 (0.08)
1D CNN	AUC	**0.82 (0.06)**	**0.95 (0.03)**
	Accuracy	**0.81 (0.04)**	**0.93 (0.01)** *
	Sensitivity	0.80 (0.05)	**0.94 (0.02)** *
	Specificity	**0.79 (0.04)**	0.88 (0.04)

In two of the thresholded metrics, performance of the 1D CNN showed statistically significant improvements over baseline on processed inputs (accuracy $= 0.93 > 0.90, p < 0.05$; sensitivity $= 0.94 > 0.89, p < 0.05$). No statistically significant difference was found for AUC ($0.95 = 0.95, p = 0.66 \nless 0.05$) and specificity ($0.88 > 0.91, p = 0.23 \nless 0.05$) across 5 repeated iterations. This then demonstrates the proposed method's ability at minimizing false negative rates with a 17.8% reduction in relative error rates while not suffering during training from class imbalance: with indistinguishable AUC metrics, we can conclude its performance is not impacted by threshold choice (Fig. 3C).

Recent studies, such as [17], have also compared SVM and CNN, showing the better performance of the latter. The researchers obtained an accuracy of 0.92 and 0.75 when using 1D CNNs with and without data augmentation, respectively. The results of the present study were obtained with data from a comparable cohort and, so far, without data augmentation in the workflow (to be included in the next stages).

Furthermore, as the proposed method improves over both SVM-based approaches—indeed no improvements were found when adding the AE component to the SVM across all metrics—we thus obtain a model that removes the need for expertly tuned features, streamlining the work of feature extraction, one of the major benefits of the DL approach. While not as relevant for use cases with already known hand-crafted features (such as this present study), these results stay nonetheless promising for generalizing to new and explored types of cancer.

When comparing the same proposed 1D CNN model trained on processed vs raw RS signals (though not accounting for the larger number of input features in the raw case), a significant improvement was observed in accuracy ($0.93 > 0.81, p < 0.05$), sensitivity ($0.94 > 0.80, p < 0.05$) and specificity ($0.88 > 0.79, p < 0.05$). This confirms that RS optical signal processing remains the gold standard for maximizing tissue discrimination results in all approaches—significant difference was found as well in AUC, accuracy and specificity for SVM, and AUC and accuracy for AE+SVM.

Interestingly, no difference in AUC, accuracy or sensitivity was found between 1D CNN and SVM which were both trained on raw inputs: we assume the noise level (as a product of a naturally weak signal and a significant background of the tissue) inherent to unadulterated RS acquisitions poses an upper bound in classification performance. Indeed, while every raw input model presented worse results in terms of accuracy, specificity and AUC when compared with the baseline SVM on processed RS signals, no statistically significant difference was found for sensitivity for our proposed method on raw signals and the (processed RS) baseline in the metric with highest clinical relevance: a sensitivity of $0.80 < 0.89, p = 0.06 \nless 0.05$. The resulting lower 20% false negative rate allows the critical improvements for margin confirmation over the current 20–35% rate during tumor removal surgery detected only postoperatively [12]. The fact that this is achieved from raw RS without manual intervention shows the potential to facilitate real-time analysis (once trained, the inference time range is 1.5 ms–2.1 ms for a single spectrum).

The major limitation of this study is the low number of patients and samples (20 and 169, respectively). Current ongoing studies are actively testing alternative wide-field RS systems to increase the number of samples and evaluating the proposed model in larger available brain RS datasets. The proposed model showed slight improvements over traditional SVMs, which could make it suitable for future integration into OR workflows, combined with recently developed open-source tools [23] or DL strategies for standardized and effective processing.

5 Conclusion

In this study, we proposed a DL pipeline based on shallow CNNs for breast cancer detection using RS that allows real-time automated feature extraction and classification. Results show improvements in accuracy and sensitivity with respect to conventional SVM-based approaches when analyzing processed RS signals and on-par performance with raw data. The fundamental role of signal processing in cancer detection was also confirmed on ex-vivo data. While results were shown to be similar to the state-of-the-art when using raw RS, there were no significant differences in false negative rates either, suggesting its potential as an alternative for incorporating RS in clinical OR workflows. This research thus allows to get closer to a real-time implementation and helps reduce the need for follow-up interventions in breast-conserving surgery. Future work will evaluate the robustness on different tissue types and the role of data augmentation.

Acknowledgements. This research was undertaken thanks, in part, to funding from the Canada First Research Excellence Fund through the TransMedTech Institute.

Disclosures. Frédéric Leblond is co-founder of ODS Medical (now Reveal Surgical) formed in 2015 to commercialize a Raman spectroscopy system for neurosurgical and prostate surgery applications. He has ownership and patents in the company.

Appendix

Figure 4 shows the results of the hyperparameter optimization. Except for the learning rate, where a value of 0.001 presented better performance than the alternatives, variations in the other parameters did not significantly affect the classification performance. Thus, for the proposed model (described in Sect. 3.3), we selected those that presented a subtle advantage; in the case of the number of epochs, the range of training time for 30 epochs was 91–102 s, while for 60 epochs, the range was 173–199 s, so the former was chosen.

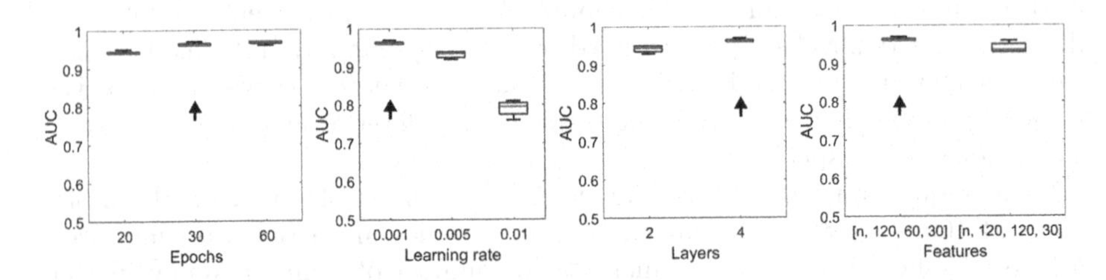

Fig. 4. AUC (mean value in red) for the 1D-CNN model during the hyperparameter optimization, testing different epochs, learning rates, number of layers, and features. Selected values are shown with an arrow. (Color figure online)

References

1. Cordero, E., Latka, I., Matthäus, C., Schie, I., et al.: In-vivo Raman spectroscopy: from basics to applications. J. Biomed. Opt. **23**(07), 1 (2018). https://doi.org/10.1117/1.jbo.23.7.071210
2. Cui, S., Zhang, S., Yue, S.: Raman spectroscopy and imaging for cancer diagnosis. J. Healthcare Eng. **2018** (2018). https://doi.org/10.1155/2018/8619342

3. Dallaire, F., et al.: Quantitative spectral quality assessment technique validated using intraoperative in vivo Raman spectroscopy measurements. J. Biomed. Opt. **25**(04), 1 (2020). https://doi.org/10.1117/1.jbo.25.4.040501
4. David, S., et al.: In situ Raman spectroscopy and machine learning unveil biomolecular alterations in invasive breast cancer. J. Biomed. Opt. **29**(03), 1–33 (2023)
5. Desroches, J., Jermyn, M., Mok, K., Lemieux-Leduc, C., et al.: Characterization of a Raman spectroscopy probe system for intraoperative brain tissue classification. Biomed. Opt. Express **6**(7), 2380 (2015). https://doi.org/10.1364/boe.6.002380
6. Desroches, J., et al.: Development and first in-human use of a Raman spectroscopy guidance system integrated with a brain biopsy needle. J. Biophotonics **12**(3), 1–7 (2019). https://doi.org/10.1002/jbio.201800396
7. Elumalai, S., Managó, S., De Luca, A.C.: Raman microscopy: progress in research on cancer cell sensing. Sensors (Switzerland) **20**(19), 1–19 (2020). https://doi.org/10.3390/s20195525
8. Gao, P., Han, B., Du, Y., Zhao, G., et al.: The clinical application of Raman spectroscopy for breast cancer detection. J. Spectrosc. **2017**(1) (2017). https://doi.org/10.1155/2017/5383948
9. Giaquinto, A.N., Sung, H., Miller, K.D., Kramer, J.L., et al.: Breast cancer statistics, 2022. CA Cancer J. Clin. **72**(6), 524–541 (2022)
10. Grajales, D., et al.: Image-guided Raman spectroscopy navigation system to improve transperineal prostate cancer detection. Part 2: in-vivo tumor-targeting using a classification model combining spectral and MRI-radiomics features. J. Biomed. Opt. **27**(09), 1–16 (2022). https://doi.org/10.1117/1.jbo.27.9.095004
11. Haka, A.S., Shafer-Peltier, K.E., Fitzmaurice, M., Crowe, J., et al.: Diagnosing breast cancer by using Raman spectroscopy. Proc. Natl. Acad. Sci. USA **102**(35), 12371–12376 (2005). https://doi.org/10.1073/pnas.0501390102
12. Jacobs, L.: Positive margins: the challenge continues for breast surgeons. Ann. Surg. Oncol. **15**(5), 1271–1272 (2008). https://doi.org/10.1245/s10434-007-9766-0
13. Kazemzadeh, M., Hisey, C.L., Martinez-calderon, M., Chamley, L.W., et al.: Deep learning as an improved method of preprocessing biomedical raman spectroscopy data, pp. 1–9 (2022). https://doi.org/10.36227/techrxiv.19435718.v1
14. Lazaro-Pacheco, D., Shaaban, A.M., Rehman, S., Rehman, I.: Raman spectroscopy of breast cancer. Appl. Spectrosc. Rev. **55**(6), 439–475 (2020). https://doi.org/10.1080/05704928.2019.1601105
15. Lemoine, É., Dallaire, F., Yadav, R., Agarwal, R., et al.: Feature engineering applied to intraoperative in vivo Raman spectroscopy sheds light on molecular processes in brain cancer: a. In: The Royal Society of Chemistry, pp. 6517–6532 (2019). https://doi.org/10.1039/c9an01144g
16. Lopes, R.M., Silveira, L., Silva, M.A.R., Leite, K.R.M., et al.: Diagnostic model based on Raman spectra of normal, hyperplasia and prostate adenocarcinoma tissues in vitro. Spectroscopy **25**(2), 89–102 (2011). https://doi.org/10.3233/SPE-2011-0494
17. Ma, D., Shang, L., Tang, J., Bao, Y., et al.: Classifying breast cancer tissue by Raman spectroscopy with one-dimensional convolutional neural network. Spectrochimica Acta - Part A: Molec. Biomolec. Spectrosc. **256**, 119732 (2021). https://doi.org/10.1016/j.saa.2021.119732
18. Pardo, A., Streeter, S.S., Maloney, B.W., et al.: Modeling and synthesis of breast cancer optical property signatures with generative models. IEEE Trans. Med. Imaging **40**(6), 1687–1701 (2021). https://doi.org/10.1109/TMI.2021.3064464
19. Paszke, A., Gross, S., Massa, F., Lerer, A., et al.: PyTorch: an imperative style, high-performance deep learning library. Adv. Neural Inf. Process. Syst. **32** (2019)

20. Petersen, D., Naveed, P., Ragheb, A., Niedieker, D., et al.: Raman fiber-optical method for colon cancer detection: cross-validation and outlier identification approach. Spectrochimica Acta - Part A **181**, 270–275 (2017). https://doi.org/10.1016/j.saa.2017.03.054
21. Plante, A., Dallaire, F., Grosset, A.A., Nguyen, T., Birlea, M., et al.: Dimensional reduction based on peak fitting of Raman micro spectroscopy data improves detection of prostate cancer in tissue specimens. J. Biomed. Opt. **26**(11), 116501 (2021). https://doi.org/10.1117/1.jbo.26.11.116501
22. Santilli, A.M., Jamzad, A., Janssen, N.N., et al.: Perioperative margin detection in basal cell carcinoma using a deep learning framework. Int. J. Comput. Assist. Radiol. Surg. **15**(5), 887–896 (2020). https://doi.org/10.1007/s11548-020-02152-9
23. Sheehy, G., Picot, F., Dallaire, F., Ember, K., Nguyen, T., Leblond, F.: Open-sourced Raman spectroscopy data processing package implementing a novel baseline removal algorithm validated from multiple datasets acquired in human tissue and biofluids. J. Biomed. Opt. **28**(February), 1–20 (2023). https://doi.org/10.1117/1.JBO.28.2.025002
24. St John, E.R., Balog, J., McKenzie, J.S., Rossi, M., et al.: Rapid evaporative ionisation mass spectrometry of electrosurgical vapours for the identification of breast pathology: Towards an intelligent knife for breast cancer surgery. Breast Cancer Res. **19**(1), 1–14 (2017). https://doi.org/10.1186/s13058-017-0845-2
25. Stomp-Agenant, M., van Dijk, T., R. Onur, A., Grimbergen, M., et al.: In vivo Raman spectroscopy for bladder cancer detection using a superficial Raman probe compared to a nonsuperficial Raman probe. J. Biophotonics **15**(6), 1–9 (2022). https://doi.org/10.1002/jbio.202100354
26. Van Rossum, G., Drake, F.: Python 3 Reference Manual. CreateSpace, Scotts Valley (2009)
27. Zhang, L., Li, C., Peng, D., Yi, X., et al.: Raman spectroscopy and machine learning for the classification of breast cancers. Spectrochimica Acta - Part A: Molec. Biomolec. Spectrosc. **264**, 120300 (2022). https://doi.org/10.1016/j.saa.2021.120300
28. Zhou, M., Hu, Y., Wang, R., Guo, T., et al.: An end-to-end deep learning approach for Raman spectroscopy classification. J. Chemom. **37**, 1–16 (2022). https://doi.org/10.1002/cem.3464

Fully Automated CAD System for Lung Cancer Detection and Classification Using 3D Residual U-Net with multi-Region Proposal Network (mRPN) in CT Images

Anum Masood[1,2](✉), Usman Naseem[3], and Mehwish Nasim[4,5,6]

[1] Department of Nuclear Medicine, RWTH University Hospital, Aachen, Germany
[2] Institute of Neuroscience and Medicine (INM-2), Forschungszentrum Jülich, Jülich, Germany
a.masood@fz-juelich.de
[3] College of Science and Engineering, James Cook University, Townsville, Australia
usman.naseem@jcu.edu.au
[4] College of Science and Engineering, Flinders University, Adelaide, Australia
[5] School of Computer and Mathematical Sciences, University of Adelaide, Adelaide, Australia
[6] School of Physics, Mathematics, and Computing, University of Western Australia, Adelaide, Australia
mehwish.nasim@uwa.edu.au

Abstract. Lung cancer is one of the leading causes of mortality worldwide. The survival rate of lung cancer depends on its timely detection and diagnosis. For pulmonary cancer detection, numerous Computer-Assisted Diagnosis (CADx) systems have been developed that use the CT scan imaging modality. Recent advancement in deep learning techniques has enabled these CADx to automatically model high-level abstractions in CT-Scan images using a multi-layered Convolutional Neural Network (CNN). Our proposed CAD system comprises 3D residual U-Net for nodule detection. Initially, the 3D residual U-Net resulted in false positive results; therefore, a multi-Region Proposal Network (mRPN) was proposed for the improvement of nodule detection. The detected nodules are assigned a probability of malignancy. Furthermore, each detected nodule is classified into four classes based on its respective malignancy score. Extensive experimental results illustrate the effectiveness of our 3D residual U-Net model. These results demonstrate the exceptional detection performance achieved by our proposed model with a sensitivity of 97.65% and an average classification accuracy of 96.37%. Performance analysis demonstrates the potential of the proposed CAD system for the detection and classification of lung nodules with high efficiency and precision.

Keywords: CT image · Lung cancer · CAD systems · Deep Learning · 3D U-Net

© The Author(s), under exclusive license to Springer Nature Switzerland AG 2023
S. Ali et al. (Eds.): CaPTion 2023, LNCS 14295, pp. 29–39, 2023.
https://doi.org/10.1007/978-3-031-45350-2_3

1 Introduction

Lung cancer has the highest mortality rate in both males and females where the 3-year survival rate for patients with lung cancer is 25% [19]. There are no obvious symptoms at the beginning of lung cancer, and as a consequence, most patients seek treatment at the later stage, minimizing survival chances. Therefore, early detection and diagnosis of lung cancer is of the utmost importance [1]. The chest computed tomography (CT) imaging modality provides high-resolution images of nodules with lavish details; however, pulmonary nodules have inhomogeneous densities and lower contrast compared to blood vessel segments and other anatomical structures, increasing the complexity of nodule detection [2,12]. To assist radiologists in automatically detecting nodules and replacing the time-consuming manual delineation of nodules, Computer-Aided Detection (CADe) and Diagnosis (CADx) systems are developed. The latest technologies use Artificial Intelligence (AI) to assist in the auxiliary diagnosis of the disease and improve the overall accuracy of the diagnosis while decreasing the detection time [10]. In recent literature, researchers have presented deep learning-based CAD systems with promising results. The convolutional neural network (CNN) framework has been used for the classification of nodules [7] and the reduction of false positive (FP) [20]. Shen et al. proposed a Multi-Crop CNN (MC-CNN) [18] and Setio et al. developed Multi-View CNN (MV-CNN) [16] to classify lung nodule. A 3D-CNN model based on Volumes of Interest (VOI) and a Fully Convolutional Network (FCN) was used to produce a score map for nodule classification [9]. Both CADe and CADx systems have been independently investigated [6], CADe are unable to provide lesion's radiological characteristics, consequently missing crucial information, while CADx systems do not identify lesions and therefore do not possess high levels of automation. Therefore, a new and advanced CAD system is needed that incorporates the benefits of detection from CADe and diagnosis from CADx into a single system for better performance.

1.1 Contribution

Our contribution is as follows:

- **3D Residual U-Net Model** A novel nodule detection method is proposed using 3D CT images for candidate nodule detection; compared to existing 2D U-Net models, our 3D residual model considers rich spatial features and therefore has more discriminative selection criteria.
- **Multi-Region Proposal Network (mRPN)** We added four RPNs so that nodules with varying diameters can be detected with ease and efficiency. The RPN split-and-merge cascade network mitigates the problem of undetected small nodules.
- **Malignancy Score-Based Approach (MSBA)** Malignancy score is calculated to classify each detected nodule into one of the four classes based on its aggregate malignancy score.

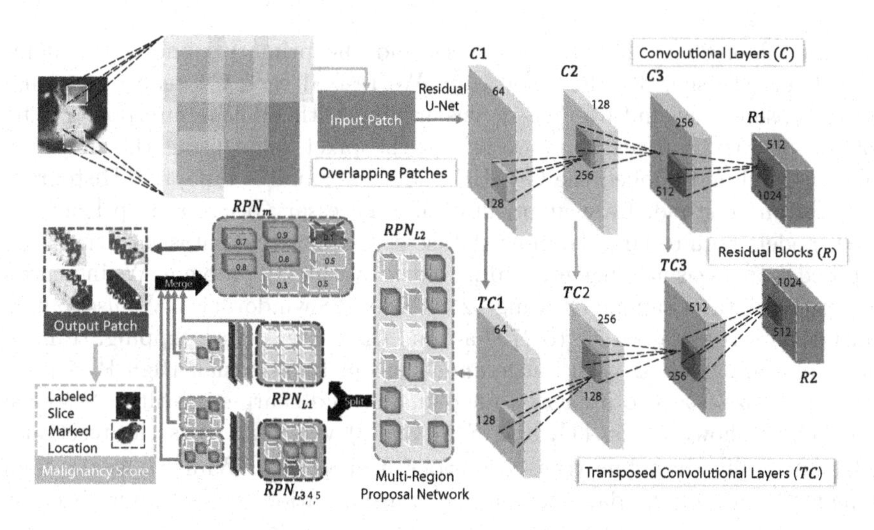

Fig. 1. Overview of our CAD system comprising of 3D residual U-Net and multi-Region Proposal Network for lung nodule detection and classification.

- **False-Positive Reduction Algorithm (FPRA)** We proposed an algorithm for a false positive (FP) reduction rate. Comparative results are much better than the existing FP reduction algorithm. The performance of the proposed CAD system is evaluated with state-of-the-art CAD systems using various performance evaluation metrics. The experimental results showed that the proposed method can not only be used for detection but also performs well for the classification of pulmonary cancer nodules as *malignant* and *benign*.

1.2 Paper Organization

Section 2 describes our approach, Sect. 3 discusses the implementation details, Sect. 4 describes the experimental results, and Sect. 5 concludes this paper.

2 Our Approach

We improved three aspects of the lung cancer detection models. First, the datasets used by most authors do not consider the lung wall. We considered location as one of the most important features and therefore found that most nodules occur in close proximity to the lung wall. Therefore, our model avoided omitting the edge of the lung. Secondly, we used 3d U-Net to filter candidate nodules [22]. Finally, we applied CNNs for nodule classification.

2.1 Pre-processing

Our in-house dataset comprises 56 patients' CT scans using a GE CT scanner (with contrast and 3mm slice thickness) in DICOM format. Each CT scan is composed of 80 to 200 distinct slices and the primary tumor was manually delineated using 3D-Slicer software. We resized each CT-scan data using average upsampling and average downsampling with bilinear interpolation by ImageJ software. Furthermore, we scaled the pixel value using the min-max scalar method and applied the CLAHE (Contrast Limited Adaptive Histogram Equalization) method. Data augmentation is necessary because deep learning-based models require large training datasets. The positive dataset to train our proposed model was insufficient, leading to the overfitting problem. We increased the positive dataset sample by using $128 \times 128 \times 128$ window size. We used affine transformations (rotation [0° to 270° around the center point], flipping, translation, and scaling) along with image enhancement by Gaussian High Pass filter with kernel size 3×3 to improve image quality and sharpening filter. For each image $P_n(\mathrm{z})$ shown in Eq. (1), and S in Eq. (2) where r_t, z_t is the reset and update to apply the affine transformations, respectively. While $\hat{h}_t$ is the final augmented state of the data. The affine transformations were standardized so that the average samples have variance=1 and mean=0.

$$P_n(\mathrm{z}) = \frac{1}{\sigma\sqrt{2\pi}} \mathrm{e}^{-\frac{(\mathrm{z}-\mu)^2}{2\sigma^2}} \tag{1}$$

$$\mathrm{S}[\mathrm{r_t}, \mathrm{z_t}] = \Sigma_{\mathrm{k}=-\mathrm{r}}^{\mathrm{n}} \mathrm{W}(\mathrm{b}, \mathrm{W} + \mathrm{hr}) \tag{2}$$

2.2 RPN Split-Merge Cascade Network

For the detection of various nodules having different diameters, we used varying levels of RPN_{Lx} referring to different sizes of the nodule. We set the RPN_{L1} as small anchors to detect the diameter of the nodule τ that ranges from 3 mm to 10 mm and has a volume $\upsilon <= 80\,\mathrm{mm}^3$ while RPN_{L2}, RPN_{L3} and RPN_{L4} have large anchors to detect nodules ranging from $\tau = 10$ mm–20 mm or $\upsilon = 80$–200 mm^3, $\tau = 20$ mm–30 mm or $\upsilon >= 200$–300 mm^3 and $\tau >= 30$ mm or $\upsilon > 300\,\mathrm{mm}^3$, respectively. The motivation behind these RPNs is the four stages of lung cancer that are categorized by different diameters, while RPN_{L1} is for all the input nodules. The RPN split and merge cascade network starts with the RPN_{L1} which is further split to either RPN_{L2} or the rest and then in the next step it is input for the RPN_{L3} or RPN_{L4}. Since each RPN level generates separate RoI sets, a merging layer is required that combines the RoI sets into one, the RPN levels merge layer RPN_m takes the input RoI sets from all the RPN levels (RPN_{L1}, RPN_{L2}, RPN_{L3}, RPN_{L4}) and outputs an aggregate RoI set RoI_{agg}. For the possibility of duplicate RoI or low objectiveness score RoI, we used the non-maxima suppression (Non-MS) when the intersection over union (IoU) overlap is above the threshold (threshold set at $\rho_t = 0.5$). After using the Non-MS, we selected the top hundred RoI with low objectiveness scores for further use. Nodule detection using different levels of RPN having various anchors improves the detection phase, as both diameter and volume are taken into consideration.

2.3 3D Residual U-Net Training Strategy and Architecture

Our proposed model relies heavily on exploiting the symmetries of the 3D space [16]. Therefore, the lung CT scan is converted into 3D fragments which are used as input for the 3D residual net. For 3D CT images containing lung nodules, the lung nodule regions were cropped to a size of $128 \times 128 \times 64$. The 3D residual net detects the module malignancy based on the characteristics obtained from the input image, and the probability of cancer stage is estimated [5]. We used binary valued threshold (corrosion & expansion) and Laplacian of Gaussian to segment the lung nodules (including lung wall), morphologic closing was performed and connected component operations are labeled in order to remove background and noise. A hole-filling algorithm based on contour information was also used to reserve the nodules on the lung wall. We obtained a collection of the interval $[x_k, y_k]$ that contains all the intervals of C_n. $\mathbf{M}$, so if n is large enough, $\sum_{k=1}^{n} |y_k - x_k| < \eta$. But $\sum_{k=1}^{n} |f_c(y_k) - f_c(x_k)| = 1$. The segmentation issue is addressed by taking N partitions of the set of features represented by P of classes M, thus minimizing the cost term of the error function by assigning the pixel P in Eq. (3) and Eq. (4)

$$\min_{\mathbf{M},\mathbf{x}} \sum_{i=1}^{N} \|\mathbf{y}_i - \mathbf{P}\mathbf{x}_i\|_2^2 \quad \text{s.t. } \forall i \ \|\mathbf{x}_i\|_0 \tag{3}$$

$$E = \frac{1}{2} \sum_{k=1}^{N} \sum_{l=1}^{M} R_{kl}^{n} V_{kl}^{2} \tag{4}$$

Another challenge for our proposed model was learning the complex inner spatial relationship between parameters using deeper CNN. We added multiple residual blocks in the middle of the 3D U-Net model, which is capable of producing higher-level packet information. Taking into account the complex anatomical structures surrounding the lung lesion, we needed an effective method to use contextual information at multiple levels [15]. An overview of the CAD system for the detection of lung nodules using the 3D residual U-Net and multi-Region Proposal Network (mRPN) is shown in Fig. 1.

2.4 Malignancy Score-Based Approach (MSBA)

For the classification of detected lung lesions in the nodule detection phase, the Malignancy Score-Based Approach (MSBA) is used to achieve the assessment of nodule malignancy of candidate lesions. For this phase, the regions of interest (RoIs) that are marked by the nodule detection phase are redefined from each marked location resulting from the last step. MSBA assigns the malignancy score to RoIs by considering the metastasis information provided in the data set to classify the candidate nodule into $T0$, $T1$, $T2$, and $T3$ stages. The details of MSBA are provided in Algorithm 1. The neighboring pixels in 3D surrounding the RoIs are taken into consideration in terms of intensity values and their eigenvalues (Hessian Matrix and the Gradient Matrix) to assign an aggregate score to each candidate-marked lesion. The result of this step is the allocation of the average score to all candidate nodules.

Algorithm 1. Malignancy Score-Based Approach

Require: $n + 1$ candidate nodules $(x_0, f(x_0)), (x_1, f(x_1)), \ldots, (x_n, f(x_n))$
Ensure: Probability of the nodule to be pulmonary cancer nodule $P^n(x)$
1: $S_{i,j} \leftarrow 0, \quad 0 \leq i, j \leq n;$
2: $S_{i,j} = f[x_{i-j}, x_{i-j+1}, \ldots, x_i];$
3: **for** $i \leftarrow 0$ to n **do**
4: $\quad S_{i,0} \leftarrow f(x_i);$
5: $\quad$ Thus, $f(x)$ is obtained by taking the singular value decomposition;
6: **end for**
7: **for** $i \leftarrow 1$ to n **do**
8: $\quad$ **for** $j \leftarrow 1$ to i **do**
9: $\quad\quad S_{i,j} \leftarrow \frac{S_{i,j-1} - S_{i-1,j-1}}{x_i - x_{i-j}};$
10: $\quad\quad$ Sum of all the prediction values is done taking candidate nodule probability;
11: $\quad$ **end for**
12: **end for**
13: $P^n(x) \leftarrow f(x_0);$
14: $R^n(x) \leftarrow 1;$
15: **for** $i \leftarrow 1$ to n **do**
16: $\quad R^n(x) \leftarrow R^n(x) \cdot (x - x_{i-1});$
17: $\quad P^n(x) \leftarrow P^n(x) + S_{i,i} \cdot R^n(x);$
18: **end for**

Table 1. Confusion matrix of lung cancer classification Using 3D Residual U-Net

Stage	T0	T1	T2	T3
T0	**96.24% (652)**	3.32% (24)	5.84% (42)	5.96% (45)
T1	5.94% (29)	**92.42% (741)**	3.76% (28)	6.64% (61)
T2	10.51% (93)	7.30% (84)	**91.35% (722)**	3.16% (22)
T3	4.45% (18)	13.18% (102)	6.14% (54)	**89.10% (709)**

3 Implementation

In addition to our in-house lung cancer CT dataset, we used publicly available datasets namely the LIDC-IDRI [4], ANODE09 [8], and LUNA16 [17] for evaluation. The probability of nodules is calculated for the nodule candidates generated by the classification model. On the basis of this probability, we mark the nodules as benign and malignant in Fig. 2. To reduce false positive results, we have proposed an algorithm that considers the probability of the candidate nodule and further improves the classification of the nodules into different stages while omitting false positive results at each stage. The details of our proposed algorithm for false positive reduction are provided in Algorithm 2.

4 Experimental Results

Our results are obtained using the concept that if a detected nodule is very close to the annotated nodule, we gain a score and the score is related to the FROC curve on sensitivity at 1/8, 1/4, 1/2, 1, 2, 4, and 8 [21]. We obtained a score of 0.974 (MAX = 1) by randomly selecting data records as the test set (excluding training and validation dataset), and an accuracy of approximately 0.997 was

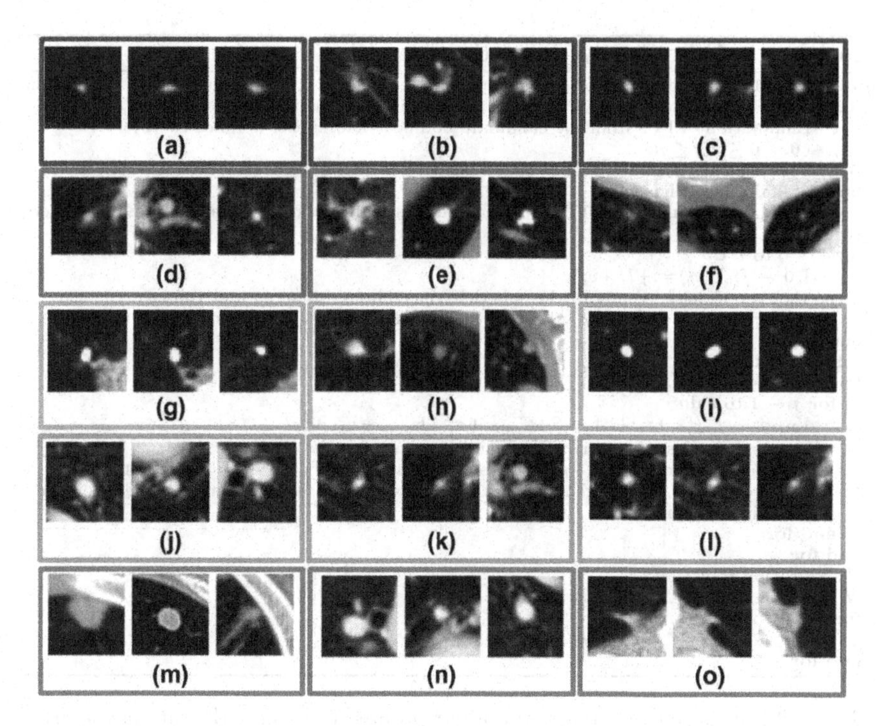

Fig. 2. Qualitative results of our proposed model on in-house dataset showing accurately detected benign nodules (T0) in (a),(b),(c); benign nodules (T1) in (d),(e),(f); malignant nodules (T2) in (g),(h),(i), malignant nodules (T3) in (j), (k) (l) and accurately detected negative nodules in (m),(n) and (o).

recorded. The results are validated using the common performance metric of computer-aided detection and diagnosis systems, i.e., average accuracy, specificity, and sensitivity. For a detailed performance analysis of our CAD system, we applied the ROC curve, which presents the TPR (True Positive Rate) as the FPR function (False Positive Rate). We plotted the sensitivity (TPR) with their respective FPR to compare our CAD system with state-of-the-art CAD systems.

The effectiveness of our method 3D Residual U-Net is verified by comparing it with Convolutional Neural Networks (CNN), Massive training artificial neural networks (MTANNs), Fully Convolutional Networks (FCN), Region-based Fully Convolutional Networks (RFCN), and RNN (Recurrent Neural Networks), the performance results are depicted in Fig. 3. Table 2 summarizes the performance of our proposed CAD system in comparison to other CAD systems in terms of accuracy, sensitivity, specificity, False Positive (FP), and average FP (using FPRA). It is visible from the confusion matrix in Table 1 that the proposed model maintains a high sensitivity for the classification of types of lung cancer even in the region where FP per scan is low. Although the proposed model achieved a high sensitivity value for stage classification for T0, T2, and T3, the sensitivity value of T1 decreases as the FP per scan becomes small, which is

Algorithm 2. False-Positive Reduction Algorithm

Require: Candidate nodule V-RoI V_R, sphericity s, ellipticity e, volume v, entropy S, maximum convergence C_{max}, intensity SD σ_I, maximum radii R_{max}, neighbors' intensity I_n
Ensure: Elimination of FPs from the candidate nodules yielding the nodules $N^p(x)$
1: $I_{i,j} \leftarrow 0, \quad 0 \leq i,j \leq n;$
2: $A_{i,j} = f[x_{i-j}, x_{i-j+1}];$
3: Initial Elimination Phase:
4: First round of Elimination of FPs from candidate nodules non-spherical and having numerous voxels;
5: **for** $i \leftarrow 0$ to n **do**
6: $V_R i, 0 \leftarrow f_{ini}(x_i){=}s{+}I_n{+}v;$
7: $f_{ini}(x)$ is obtained by sum of the sphericity, neighbors' intensity and volume;
8: If $f_{ini}(x_i)$ for any V-RoI $V_R i, 0$ is less than the threshold T_{ini}, V-RoI is eliminated;
9: **end for**
10: Advance Elimination Phase:
11: **for** $i \leftarrow 1$ to n **do**
12: **for** $j \leftarrow 1$ to i **do**
13: $V_R i, j \leftarrow f_{ad}(x_i){=}v{+}s{+}e{+}S{+}C_{max}{+}\sigma_I{+}R_{max}{+}I_n;$
14: Eigenvalues Hessian matrix $H^i(x)$ and Gradient matrix $G^i(x)$;
15: If $f_{ad}(x_i)$ for any V-RoI $V_R i, j$ is less than threshold T_{ad}, V-RoI is eliminated;
16: Features sum $f_{ad}(x)$ is done taking candidate nodule features into consideration for the classification phase;
17: **end for**
18: **end for**
19: Classification Phase:
20: $N^p(x) \leftarrow f_{cl}(x_0);$
21: **for** $i \leftarrow 1$ to n **do**
22: $N^p(x) \leftarrow N^p(i){+}f_{ad}(x_i).\ H^n(x) + G^i(x);$
23: **end for**

Table 2. Performance comparison of our proposed model with state-of-the-art

Method	Accuracy	Sensitivity	Specificity	FP per scan	Avg FP (using FPRA)
CNN [23]	0.797	0.753	0.865	0.070	0.065
TumorNet [11]	0.811	0.815	0.899	0.090	0.083
FCNN [3]	0.843	0.837	**0.967**	0.110	0.071
DFCNet [13]	0.967	0.821	0.954	**0.040**	0.029
mRFCN [14]	0.913	0.731	0.864	0.063	0.034
Ours	**0.997**	**0.976**	0.942	0.045	**0.026**

impractical in the clinical environment [14]. A summary of our proposed CAD system in terms of stage classification is shown in Table 1. The experimental results demonstrate the superiority in the classification and class generalization of our proposed 3D Residual U-Net based CAD system. A comparison of nodule classification by CAD systems is shown in Fig. 4.

A comparison of our model with the existing state-of-the-art CADe systems on the LIDC-IDRI dataset with varying nodule sizes is shown in Table 3. We have compared the detection accuracy of our CADe system with the detection accuracy of five other existing systems, which are evaluated on the dataset from the LIDC-IDRI database. The high accuracy of our proposed model with varying sizes signifies the detection capability of our 3D Residual U-Net model. Our method has promising results in discriminating cancer nodule types without compromising detection accuracy (Fig. 4).

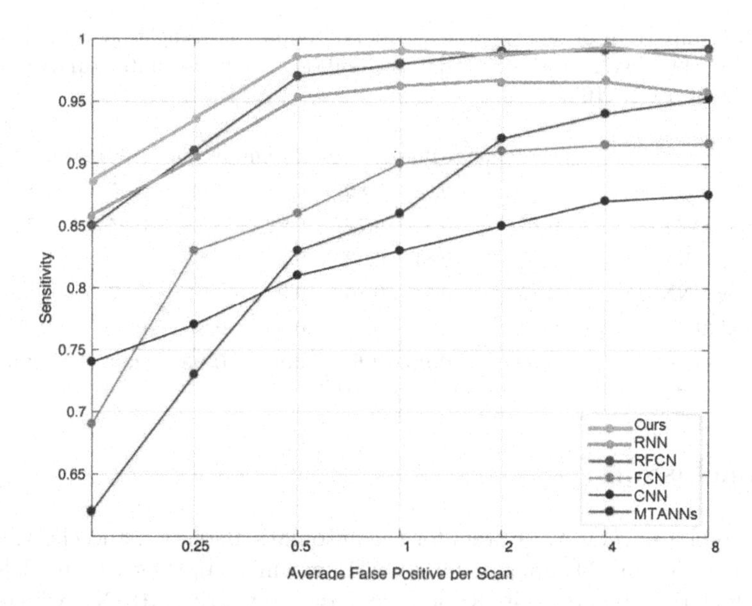

Fig. 3. The plot illustrating the sensitivity of the 3D residual model on the Luna16 dataset with the state-of-the-art CAD systems. The x-axis denotes the average number of FP per scan while the y-axis represents the sensitivity.

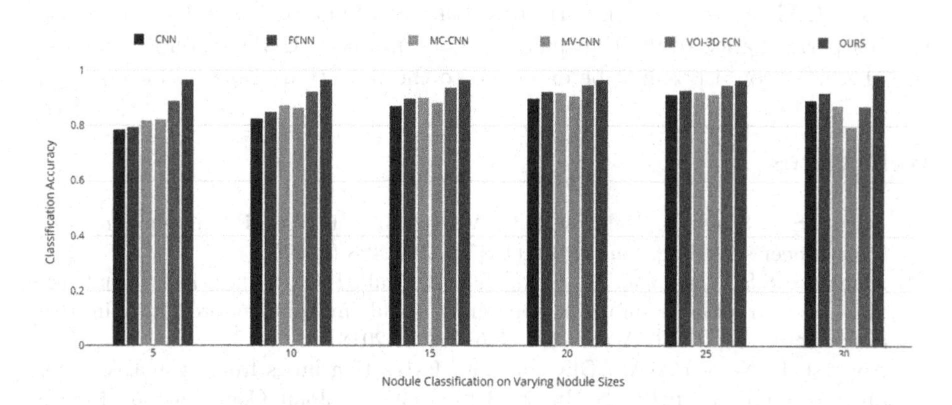

Fig. 4. Comparison of our 3D residual U-Net CAD system nodule classification with state-of-the-art CADe systems.

Table 3. Comparison of our proposed model on nodule classification with the state-of-the-art CADe systems using CT dataset (subset of inhouse data, LIDC-IDRI [4], ANODE09 [8], LUNA16)

Model	<= 3 mm	5 mm	10 mm	20 mm	30 mm	>30 mm	Avg ACC*
CNN [20]	0.67	0.78	0.86	0.91	0.88	0.85	0.83
FCNN and SVM [7]	0.75	0.79	0.77	0.91	0.92	0.93	0.73
Multi-Crop CNN [18]	0.68	0.81	0.90	0.89	0.87	0.91	0.85
Multi-View CNN [16]	0.67	0.81	0.86	0.90	0.91	0.87	0.84
VOI Based 3D-FCN [9]	0.90	0.88	0.93	0.94	0.87	0.96	0.91
Ours(3D Residual U-Net)	**0.94**	**0.96**	**0.96**	**0.97**	**0.98**	**0.98**	**0.96**

* Avg ACC means Average Accuracy; Row 2 to Row 7 represent different nodule sizes

5 Conclusion

We proposed a novel CAD system for the automatic detection and classification of lung nodules in CT images. Our CAD system comprises two models, the 3D residual U-Net and multi-Region Proposal Network (mRPN), which have demonstrated effective nodule detection results even with small-sized lesions. For the classification of nodules, the aggregate malignancy score is calculated for each detected nodule. Based on this score, detected nodules are classified into four classes; $T0$, $T1$, $T2$, and $T3$. Experimental results illustrate the efficacy of our proposed CAD system in comparison to state-of-the-art CAD systems that use various performance evaluation metrics. Our proposed CAD system is generic and therefore could possibly be extended to the detection of other cancers.

References

1. Adams, S.J., Stone, E., Baldwin, D.R., Vliegenthart, R., Lee, P., Fintelmann, F.J.: Lung cancer screening. Lancet **401**(10374), 390–408 (2023)
2. Alahmari, S.S., Cherezov, D., Goldgof, D.B., Hall, L.O., Gillies, R.J., Schabath, M.B.: Delta radiomics improves pulmonary nodule malignancy prediction in lung cancer screening. IEEE Access **6**, 77796–77806 (2018)
3. Alves, J.H., Neto, P.M.M., Oliveira, L.F.: Extracting lungs from ct images using fully convolutional networks. In: 2018 International Joint Conference on Neural Networks (IJCNN), pp. 1–8. IEEE (2018)
4. Armato, S.G., III.: The lung image database consortium (LIDC) and image database resource initiative (IDRI): a completed reference database of lung nodules on CT scans. Med. Phys. **38**(2), 915–931 (2011)
5. Dou, Q., Chen, H., Yu, L., Qin, J., Heng, P.A.: Multilevel contextual 3-D CNNs for false positive reduction in pulmonary nodule detection. IEEE Trans. Biomed. Eng. **64**(7), 1558–1567 (2017)
6. Firmino, M., Angelo, G., Morais, H., Dantas, M.R., Valentim, R.: Computer-aided detection (CADe) and diagnosis (CADx) system for lung cancer with likelihood of malignancy. BioMed. Eng. OnLine **15**(1), 2:1–2:17 (2016)

7. van Ginneken, B., Setio, A.A.A., Jacobs, C., Ciompi, F.: Off-the-shelf convolutional neural network features for pulmonary nodule detection in computed tomography scans. In: IEEE International Symposium on Biomedical Imaging, pp. 286–289 (2015)
8. van Ginneken, B., et al.: Comparing and combining algorithms for computer-aided detection of pulmonary nodules in computed tomography scans: the ANODE09 study. Med. Image Anal. **14**(6), 707–722 (2010)
9. Hamidian, S., Sahiner, B., Petrick, N., Pezeshk, A.: 3D convolutional neural network for automatic detection of lung nodules in chest CT. In: Proceedings of SPIE, vol. 10134 (2017)
10. Huang, S., Yang, J., Shen, N., Xu, Q., Zhao, Q.: Artificial intelligence in lung cancer diagnosis and prognosis: current application and future perspective. In: Seminars in Cancer Biology. Elsevier (2023)
11. Hussein, S., Gillies, R., Cao, K., Song, Q., Bagci, U.: TumorNet: lung nodule characterization using multi-view convolutional neural network with Gaussian process. In: IEEE International Symposium on Biomedical Imaging, pp. 1007–1010 (2017)
12. Li, X., Deng, Z., Deng, Q., Zhang, L., Niu, T., Kuang, Y.: A novel deep learning framework for internal gross target volume definition from 4d computed tomography of lung cancer patients. IEEE Access **6**, 37775–37783 (2018)
13. Masood, A., et al.: Computer-assisted decision support system in pulmonary cancer detection and stage classification on CT images. J. Biomed. Inf. **79**, 117–128 (2018)
14. Masood, A., et al.: Automated decision support system for lung cancer detection and classification via enhanced rfcn with multilayer fusion rpn. IEEE Trans. Ind. Inf. **16**(12), 7791–7801 (2020)
15. Masood, A., et al.: Cloud-based automated clinical decision support system for detection and diagnosis of lung cancer in chest CT. IEEE J. Transl. Eng. Health Med. **8**, 1–13 (2019)
16. Setio, A.A.A., et al.: Pulmonary nodule detection in CT images: false positive reduction using multi-view convolutional networks. IEEE Trans. Med. Imaging **35**(5), 1160–1169 (2016)
17. Setio, A.A.A., et al.: Validation, comparison, and combination of algorithms for automatic detection of pulmonary nodules in computed tomography images: The LUNA16 challenge. Med. Image Anal. **42**, 1–13 (2017)
18. Shen, W., et al.: Multi-crop convolutional neural networks for lung nodule malignancy suspiciousness classification. Pattern Recogn. **61**, 663–673 (2017)
19. Siegel, R.L., Miller, K.D., Wagle, N.S., Jemal, A.: Cancer statistics, 2023. Ca Canc. J. Clin. **73**(1), 17–48 (2023)
20. Tan, M., Deklerck, R., Jansen, B., Bister, M., Cornelis, J.: A novel computer-aided lung nodule detection system for CT images. Med. Phys. **38**(10), 5630–5645 (2011)
21. Teramoto, A., Fujita, H.: Fast lung nodule detection in chest CT images using cylindrical nodule-enhancement filter. Int. J. Comput. Assist. Radiol. Surg. **8**(2), 193–205 (2013)
22. Xie, Y., Zhang, J., Xia, Y., Fulham, M., Zhang, Y.: Fusing texture, shape and deep model-learned information at decision level for automated classification of lung nodules on chest CT. Inf. Fusion **42**, 102–110 (2018)
23. Yuan, J., Liu, X., Hou, F., Qin, H., Hao, A.: Hybrid-feature-guided lung nodule type classification on CT images. Comput. Graph. **70**, 288–299 (2018)

Image Captioning for Automated Grading and Understanding of Ulcerative Colitis

Flor Helena Valencia[1], Daniel Flores-Araiza[1], Obed Cerda[2], Venkataraman Subramanian[3], Thomas de Lange[4], Gilberto Ochoa-Ruiz[1], and Sharib Ali[5](✉)

[1] School of Engineering and Sciences, Tecnologico de Monterrey, Monterrey, Mexico
[2] CINVESTAV Unidad Guadalajara, Zapopan, Mexico
[3] Leeds Teaching Hospitals NHS Trust, Leeds, UK
[4] Sahlgrenska University Hospital, Gothenburg, Sweden
[5] School of Computing, University of Leeds, Leeds, UK
s.s.ali@leeds.ac.uk

Abstract. Ulcerative colitis (UC) is a chronic inflammatory disease of the large bowel characterised by quisent periodes and relapses. Endoscopic grading of the severity of UC is done by using a widely accepted scoring system known as the "Mayo Endoscopic Scoring" (MES). The MES score is largely based on the recognition of phenotypic features of the mucosal wall, and thus the subjectivity in clinical scoring is unavoidable. An automated grading and characterisation can certainly help to minimise the inter-observer variability and help trainees to get useful insights. For the first time, we a system capable of not only providing an automated MES scoring system, but also of generating a description of visible MES phenotypic mucosal representations in these endoscopic images through captions. Our aim is to combine the visual features together with word sequence embeddings that are learnt jointly through a recurrent neural network to predict such scene descriptions. In this work, we explore various recurrent neural network architectures together with other backbone architectures for visual feature representations. Our experiments on held-out test samples demonstrate high similarity between the reference and the predicted captions.

Keywords: Colonoscopy · Ulcerative colitis · Image captioning · Classification · Deep learning

1 Introduction

Inflammatory bowel diseases (IBD), including Crohn's disease (CD) and ulcerative colitis (UC), and characterised by a chronic inflammation affecting the gastrointestinal tract. As per recent estimates, the annual incidence of UC in Europe is approximately 24.3 cases per 100,000 populations per year, and in terms of prevalence it is about 322 cases per 100,000 population per year [9]. However, these numbers are projected to rise as specific geographical patterns

© The Author(s), under exclusive license to Springer Nature Switzerland AG 2023
S. Ali et al. (Eds.): CaPTion 2023, LNCS 14295, pp. 40–51, 2023.
https://doi.org/10.1007/978-3-031-45350-2_4

have indicated an increasing incidence of UC in countries such as Asia and North Africa [14]. Since UC is a lifelong illness, patients with this condition often experience relapses and remissions which lead to significant impact on their quality of life [4]. Additionally, due to the persistent nature of UC, patients have six times increased risk of colorectal cancer (CRC). Several studies indicate that the colonoscopy surveillance plays a crucial role in mitigating the development to colorectal cancer (CRC) [10]. Therefore, an early and accurate diagnosis is critical to enhance the patient's chances of long term disease remission, as treatment approaches and subsequent monitoring vary depending on the severity of the disease.

Colonoscopy is considered as the gold standard diagnostic procedure for UC as it enables an accurate assessment of the extent and severity of the disease. Several evaluation methods exist, with the Mayo Endoscopic Score (MES) being the most extensively used [5,11]. The MES system takes into account several features including loss of vascular pattern, erythema, friability, granularity, erosions, mucosal bleeding, and ulcerations [11,16]. Here, friability is an important characteristics that can only can be assessed during endoscopy. Based on these criteria, UC is classified into four severity scores, ranging from 0 to 3, with MES 0 indicating inactive disease with normal mucosa, MES 1 mild disease, MES 2 moderate disease, and MES 3 severe disease. However, the grading of UC in endoscopy relies mainly on the level of expertise and ability of health professionals. Similarly, UC grading can be very challenging as several images can have bleeding cause by biopsy and not be due the inflammation. Thus, it is often difficult to assess vascular pattern on one simple image because there is a variation between individuals and regions of the colon. Several studies have highlighted substantial inter- and intra-observer variability in the diagnosis and scoring of UC due to the subjectivity of qualitative image interpretation in endoscopy [15,17].

The importance of developing AI tools for automated classification of UC has increased over recent years in an attempt to improve diagnostic accuracy and reduce subjectivity. Most of these studies have employed deep learning models for developing computer-assisted diagnosis (CAD) for classifying inflammation severity according to MES [6,12,13,19]. A recent survey paper that highlighted current works on applying deep learning (DL) for UC classification suggested that UC is highly complex, with highly subtle changes between mild to severe types makes DL method struggle in accurately classifying [1]. Using multi-modal data (e.g., both images and text embedding) can help learn clear representations compared to a single modality (e.g., only images). Moreover one of the limitations of utilising DL algorithms for diagnostic tasks is the inherent black box nature of these methods, as they fail to explain the machine-generated decisions on complex findings such as UC grading, hindering clinical adoption.

In this paper we propose a proof-of-concept approach for image captioning to develop trustworthy and reliable automated scoring system for UC (Fig. 1). Our approach aims to provide comprehensive description of UC manifestations in endoscopic images. Our approach employs an Encoder-Decoder architecture. For

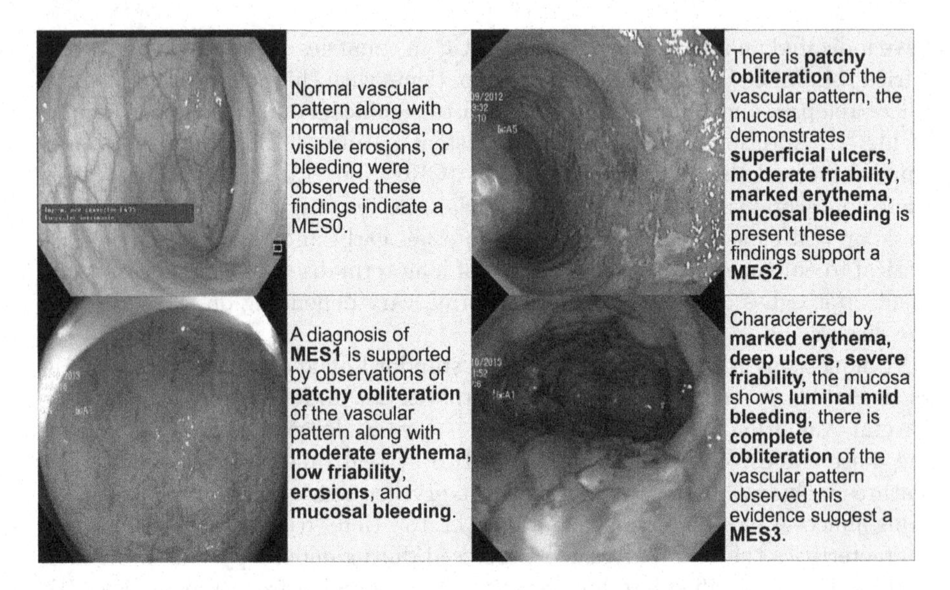

Fig. 1. Sample images with caption. Mayo Endoscopic Scoring with description as caption for each categories for grading patients with ulcerative colitis in inflammatory bowel disease.

the encoder we test a range of distinct backbone architectures, each optimised for feature extraction. For the decoder, we make use of several recurrent units including Long Short-Term Memory (LSTM), Gated Recurrent Unit (GRU), and basic Recurrent Neural Networks (RNN). The purpose of our approach is to generate accurate captions based on several endoscopic features, thereby facilitating the evaluation of disease severity based on the MES system, while tackling the prevailing issue of interpretability associated with machine-generated decisions.

The rest of this paper is organised as follows. In Sect. 2 we discuss recent relevant research on UC scoring and automated medical report generation using DL. Section 3 outlines the proposed method, while Sect. 4 covers implementation details, dataset preparation, and results. Finally, discussion and conclusion are presented in Sect. 5.

2 Related Work

2.1 Deep-Learning Based UC Scoring

In UC grading, several studies have developed deep-learning based classifications models for assessing colonoscopic inflammations. Sutton *et al.* [13] used a DenseNet121 achieving an accuracy of 87.5% and an AUC of 0.90 on the Hyper-Kvasir dataset. While Ozawa *et al.* [8] developed a GoogLeNet-based computer-aided diagnosis system trained using 26,304 colonoscopy images from a total of 841 patients, obtaining an AUROC of 0.98 for Mayo 0–1 versus 2–3 inflammations. However, these two studies were limited to a binary classification approach.

In contrast, Bhambhvani and Zamora [2] developed a model capable of individual MES discrimination (1,2 and 3) using ResNeXt-101. Their proposed model achieved AUC values of 0.96, 0.89, and 0.86 for MES 3, 2, and 1 respectively, and an overall accuracy of 77.2%. Following this similar approach, Xu *et al.* [18] proposed a 3-way MES classification model based on the EfficientNet-b5 architecture using HyperKvasir dataset. The model incorporated ArcFace loss, an additive angular margin penalty based on softmax, and achieved top-1 accuracy of 75.06%, top-2 accuracy of 93.68%, and an F1 score of 76.15%. All of these methods are only classification methods and do not use image captioning strategy to describe the endoscopic findings.

2.2 Automated Medical Report Generation

So far, a limited number of studies have been conducted related to image captioning in the context of endoscopic diagnosis. Fonollà *et al.* [6] developed a CADx system to automatically generate colorectal polyp (CRP) reports based on Blue light imaging Adenoma Serrated International Classification (BASIC) using four descriptors. The study model consists of an EfficientNetB4 based encoder and feature extractor, with the classification layers replaced by a global average pooling layer. The authors employed a pre-trained BERT module for learning polyp sequences. The visual and text information were later concatenated and passed through an LSTM to capture word temporal relations. Evaluated using n-gram based metrics like BLEU and ROUGE, and METEOR scores were used with the model providing BLUE-1 of 0.67, ROUGE-L of 0.83, and METEOR of 0.50. To the best of our knowledge, no documented research has focused on automating the report generation for ulcerative colitis.

3 Method

Our proposed method integrates the Encoder-Decoder framework (Fig. 2). We adopted different backbones for encoding and recurrent units for decoding in this architecture including DenseNet121, ResNet50, and Res2Net50 as encoders, each pre-trained on ImageNet. For the decoding part, we experimented with LSTM, GRU, and RNN and also with one or two layers.

As shown in Fig. 2, the high-level feature vectors obtained from the encoder are concatenated with the embeddings of reference captions, produced through an embedding layer, and then fed into the selected recurrent unit. To prevent overfitting and to promote model robustness, we implemented a dropout layer before the final fully-connected layer. This layer maps the outputs to the size of the vocabulary, and the resulting label indexes are subsequently converted into corresponding words, forming the predicted captions.

3.1 Feature Extraction

The initial stage of our medical image captioning model involves feature extraction. For this task, we experimented with three distinct ImageNet pre-trained

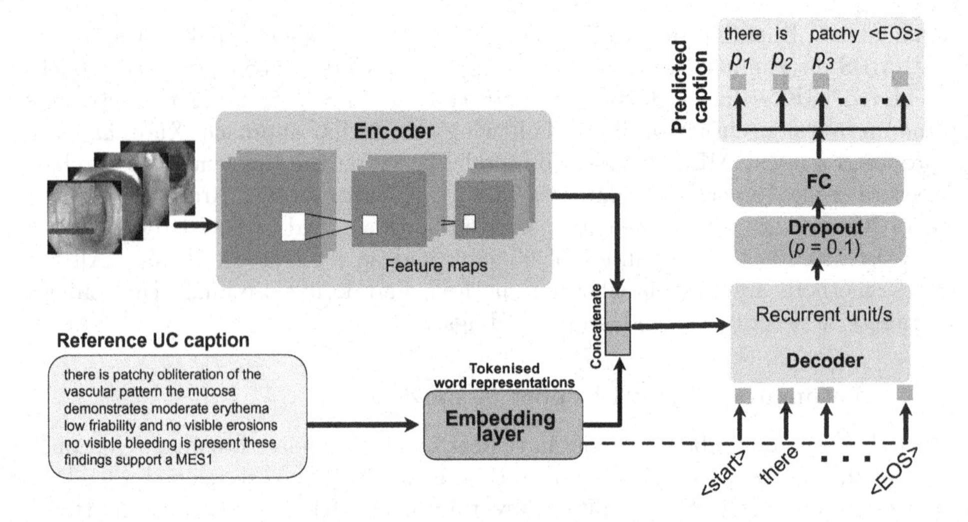

Fig. 2. UC image captioning block diagram. Feature map predictions from an ImageNet pre-trained encoder network are used and concatenated with the embeddings of the reference UC captions. Embeddings are then passed to recurrent unit (either single or two in our experiments). Finally, a dropout and fully connected layer is used for predicting label indexes that are converted to the corresponding words representing predicted caption.

Table 1. Experimental results on validation set for classification of MES (0, 1, 2 and 3). Best results is presented in bold.

Backbone	Metrics			
	Accuracy	F1-score	PPV	Recall
DenseNet121	75.75%	74.70%	76.33%	75.75%
ResNet50	75.75%	74.16%	76.74%	75.75%
Res2Net50	**78.78%**	**78.32%**	**78.08%**	**78.78%**

backbones: DenseNet121, ResNet50, and Res2Net50. We first trained these architectures to classify images based the MES into four classes: MES-0, MES-1, MES-2, and MES-3. Table 1 presents the classification accuracies corresponding to each of the selected backbones in our experimental evaluation. Once trained, the learned weights from this classification task were retained. During the encoder phase of our captioning model, images are processed by one of these pre-trained backbones, each serving as a feature extractor. This was accomplished by removing the final layer of the architecture and utilizing the weights from the MES classification training to ensure the extracted features were relevant and fine-tuned to our medical dataset.

The backbone transforms the input image into a set of high-level features, which are output of the encoder phase. To ensure stable learning, we apply batch normalization to prevent shifts in feature value distributions during training, thus promoting consistent learning progress and predictable model performance.

3.2 Sequence Processing and Decoding

For the sequential data pre-processing, the input captions were tokenized and transformed into a dense vector representation via word embedding layer. This representation reduces the input data dimensionality and captures semantic relationships between words. Within each processing batch, a reference caption was randomly selected from the set to ensure variation and enrich the training process. During this process, special tokens are introduced to handle specific scenarios, which will be described as follows.

First of all, the <SOS> (Start Of Sentence) token is used to mark the beginning of each caption, whereas the <EOS> (End Of Sentence) token signifies the conclusion of each caption. Additionally, in order to ensure consistency in the input size of the recurrent units, all captions were padded with the special token (<PAD>) in order to reach the maximum sequence length, which in this case is set to 33 words.

The resultant embeddings are subsequently concatenated with the visual feature vectors thus integrating the multi-modal information. This process produces a comprehensive representation that captures both visual features and the corresponding text annotations. The combined tensor is subsequently fed to the chosen recurrent unit. The final fully connected layer modifies the recurrent unit output to it align with the dimension of the target vocabulary. A linear transformation is used to generate the next word in the sequence by identifying the word with the highest score.

To generate captions for input images during inference, we use a greedy decoding approach. The process is initiated with the special token <SOS>. At each time step, the model generates a probability distribution over the target vocabulary for the next word, based on the current state and input. The word with the highest probability is selected as the next word in the sequence. This sequence of operations continues until the special token <EOS> token is predicted.

4 Experiments and Results

4.1 Implementation Details

Dataset The dataset used to conduct the experiments is composed of 982 images sourced from both publicly available and in-house data (only MES 0). In house data was used only to balance the class imbalance in the MES 0 category. The public Hyperkvasir dataset comprises of 851 images [3] and in-house comprised of 131 images of MES 0. The available dataset included MES scores (0 to 3); however, it lacked descriptions. We employed the widely used clinically accepted MES scoring description [11,16]. For experiments, we split the images and the corresponding caption labels into 80% for training (785), 10% for validation (99), and 10% for testing (98) purposes. Table 2 shows the sample distribution across sets based on the MES, and Fig. 3 illustrates the number of samples for the endoscopic visual feature descriptors.

Table 2. Size of train, validation and test set for each MES scoring.

MES	Train	Validation	Test
0	177	22	22
1	145	18	18
2	362	46	45
3	101	13	13

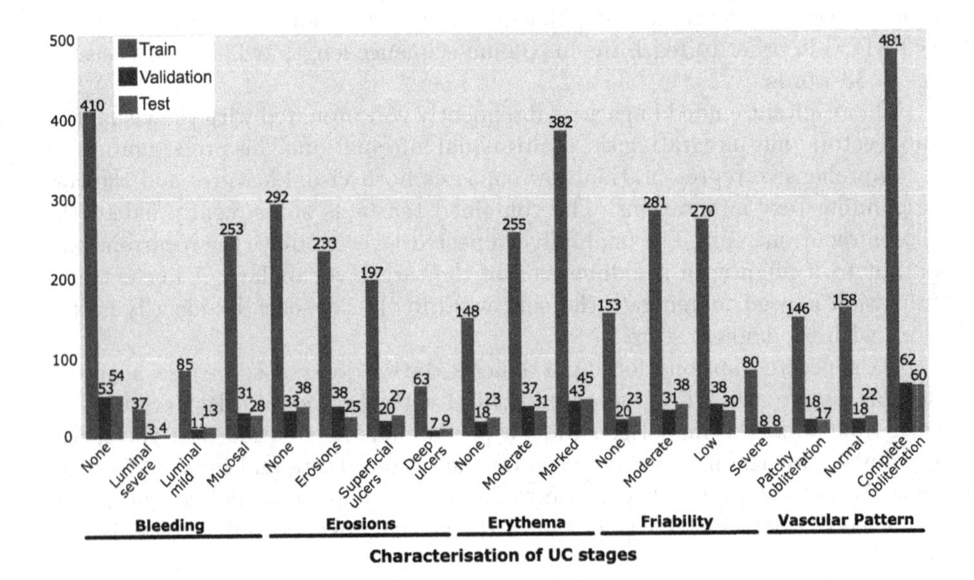

Fig. 3. Data distribution. Number of samples for each of the descriptors that are used to categorise different stages of ulcerative colitis.

Protocol for Reference Caption Generation. All labels were conducted by a the first author under the supervision of experts in the domain. Subsequently, these annotations were reviewed and adjusted by a senior gastroenterologist. Each image was associated with three distinct reference captions. While the finalised captions are derived from these expert-reviewed annotations, they do adhere to a schema. Specifically, the descriptions incorporate information from five critical endoscopic features: vascular pattern, erythema, friability, the presence of erosions or ulcers, and the MES. It is important to emphasize that these captions are dynamic and vary across images, even when they belong to the same MES grade. For instance, an MES-2 does not necessarily imply the presence of other tissue features associated with disease severity.

Example of Sample Template: *There is [vascular pattern], the mucosa demonstrates [degree of erythema], [degree of friability], and [erosion/ulcers], [degree of bleeding] is present; these findings support a [MES scoring].* It is to be noted that the template is not static and can differ depending on the mucosal properties. The same MES grade can have a very different descriptive analysis based on the mucosa.Figure 1 shows examples of the reference captions.

Training Setup. We train our UC image captioning model using the PyTorch framework on a GPU setup comprising four NVIDIA Tesla K80 GPUs. We use a learning rate (lr) of $3e^{-4}$ and an Adam optimiser. We set 100 epochs with a batch size of 32 to train the model presented in all experiments. Finally, all input images were resized to 224×224 pixels, and we use a 256-dimensional word embedding.

Evaluation Metrics. To assess the performance of our classification backbones used in this dataset, we employed widely used F1-score, positive predictive value (PPV), recall and accuracy. Finally, to assess the performance of our final model, we used a set of evaluation metrics commonly employed in image captioning tasks. These metrics provide a quantitative measure of how well a model generates accurate and semantically relevant captions. These include the BLEU (Bilingual Evaluation Understudy) using four different n-grams, ROUGE (Recall-Oriented Understudy for Gisting Evaluation) score which is recall focused metric, and Cosine similarity which measures the similarity between two texts based on the cosine of the angle between their vector representations. We also provide, frames-per second to provide the relative inference time for each model under investigation.

4.2 Evaluation Results

Quantitative Results. As observed in Table 3, GRU and RNN with DenseNet121 provided best metric values over other combinations for two-layered recurrent networks, achieving a BLEU-4 of 0.7352 and 0.7326, respectively. For example, the models which used DenseNet121 as feature extraction consistently outperformed those using ResNet50 and Res2Net50, with average improvements in BLEU-4 scores by 31.9% and 30.2%, respectively. Similarly, models employing GRU as the decoder significantly surpassed those using RNN and LSTM. On average, the GRU-based models improved the BLEU-4 scores by 3.2% and 4.2% over models using GRU and LSTM, respectively.

Qualitative Results. Figure 5 presents examples of the generated captions for each MES class. The captions are analysed for descriptor identification accuracy, wherein red highlights indicate incorrectly identified descriptors, while blue highlights denote correctly predicted descriptors. Finally, the black words in the captions correspond to related terms that are contextually expected and hold no specific relevance to the descriptor identification task.

As it can be observed in the figure, the model consistently achieved accurate predictions for the majority of the desired descriptors in each case, particularly

Table 3. Experimental results on test set for automated image description generation. The two best results are presented in bold.

Layers	Backbone	RU	Metrics						
			BLUE-1	BLUE-2	BLUE-3	BLUE-4	RL	CSim	FPS
N = 1	DenseNet121	LSTM	0.8789	0.8166	0.765	0.7203	0.8430	0.8907	22.52
		GRU	0.8768	0.8133	0.7607	0.7179	0.8430	0.8817	22.86
		RNN	0.8766	0.8117	0.7593	0.7153	0.8365	0.8863	22.95
	Res2Net50	LSTM	0.772	0.6784	0.6057	0.5491	0.7314	0.7823	33.55
		GRU	0.7782	0.6948	0.6308	0.5753	0.7405	0.7944	27.18
		RNN	0.6948	0.6741	0.6167	0.5616	0.7087	0.7635	27.88
	ResNet50	LSTM	0.6308	0.6590	0.5853	0.5357	0.7205	0.7703	32.71
		GRU	0.5753	0.6699	0.6047	0.5541	0.7178	0.7766	33.55
		RNN	0.7405	0.6695	0.6103	0.558	0.7087	0.7519	35.84
N = 2	DenseNet121	LSTM	0.7944	0.8195	0.769	0.7251	0.8911	0.8511	22.28
		GRU	**0.8855**	**0.8263**	**0.7783**	**0.7352**	**0.8981**	**0.8478**	21.93
		RNN	**0.8847**	**0.8255**	**0.7766**	**0.7326**	**0.8952**	**0.8512**	21.00
	Res2Net50	LSTM	0.7614	0.6592	0.5843	0.5354	0.7222	0.7222	24.54
		GRU	0.7760	0.6900	0.6336	0.5807	0.7205	0.7689	25.78
		RNN	0.7632	0.6619	0.5870	0.5371	0.723	0.7706	27.90
	ResNet50	LSTM	0.7614	0.6592	0.5843	0.5354	0.7222	0.7222	24.54
		GRU	0.7550	0.6641	0.6073	0.5534	0.7618	0.7058	31.39
		RNN	0.7623	0.6650	0.5969	0.5476	0.7190	0.7638	33.47

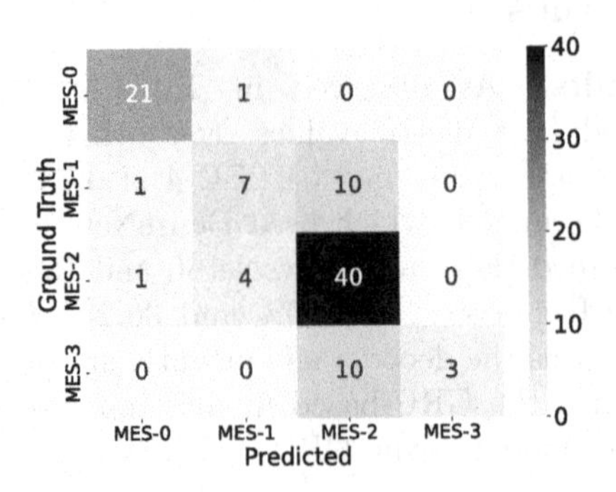

Fig. 4. Confusion matrix for our best model (DenseNet121-GRU with 2-layers)

from MES-0 to MES-2. In the case of MES-3 examples, the model encountered challenges in accurately predicting certain instances, such as distinguishing between moderate and severe friability or differentiating between superficial and deep ulcers. These specific endoscopic features present inherent challenges in discriminating between varying levels of severity.

5 Discussion and Conclusion

In this study, we aimed to investigate, the performance of various combinations of feature extraction backbones and recurrent units to generated automated descriptions for UC scoring. Our main objective was to identify the most effective combination that would produce the most accurate captions based on our descriptors. We observed that employing the DenseNet121 backbone in conjunction with 2 layers GRU and RNN obtained overall best performance, with respective BLEU-4 scores of 0.7352 and 0.7326 for automated UC scoring descriptions

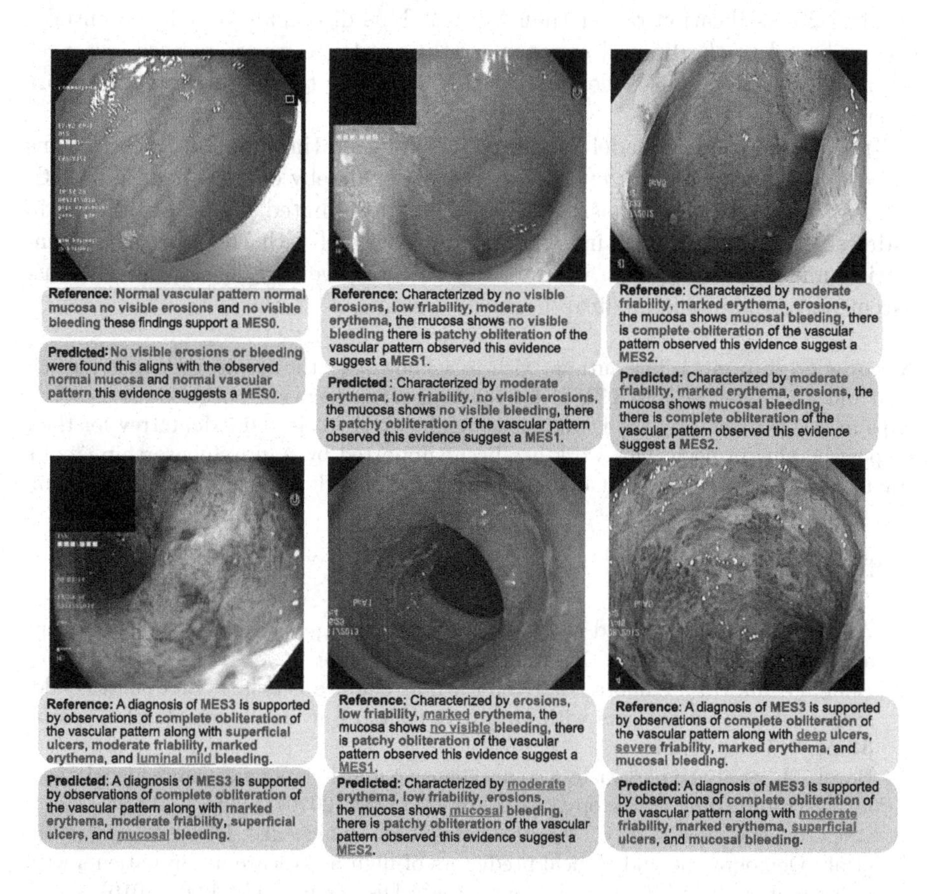

Fig. 5. Examples of generated captions for each MES class using the best model (DenseNet121 with GRU).

(Table 3). This superior performance could be attributed to several factors: 1) DenseNet121, as a feature extractor, has shown to be particularly effective for similar tasks [7] and domain [13]; 2) The dense connections and feature reuse mechanism employed by this feature extractor contribute to enhanced gradient flow and feature propagation across the network resulting in more diversified features and less redundant parameters; 3) finally as for the decoder model, despite its simplicity, the GRU effectively captured the necessary temporal dependencies in the captioning task since GRU focuses on short-term dependencies, which was suitable given the short sequence length of our captions.

The generated captions exhibited a strong similarity between the ground truth captions. Moreover, our model consistently achieved accurate prediction of severity levels for most of the endoscopic features, thus effectively grading UC in the MES-0 to MES-2 classes (Fig. 4). It is important to note that the model encountered difficulties when predicting the MES-3 classification, as well as the associated characteristics indicative of this level of severity. For this classification, the model predicted features that were more commonly associated with an MES-2 classification rather than MES-3. This discrepancy could potentially be attributed to the limited exposure of the model to severe disease instances during the training phase, thereby affecting its ability to accurately predict more severe cases.

In conclusion, our model was capable of generating accurate descriptions based on the most relevant endoscopic features, thereby effectively grading UC. Moreover, our study emphasized the value of automated report generation in addressing the lack of explainability often associated with DL models. By generating understandable and accurate descriptions, we can develop more transparent and explainable medical diagnosis systems.

Acknowledgements. The authors wish to acknowledge the Mexican Council for Science and Technology (CONACYT) for the support in terms of postgraduate scholarships in this project, and the Data Science Hub at Tecnologico de Monterrey for their support on this project. This work has been supported by Azure Sponsorship credits granted by Microsoft's AI for Good Research Lab through the AI for Health program.

References

1. Ali, S.: Where do we stand in AI for endoscopic image analysis? deciphering gaps and future directions. npj Dig. Med. **5**(1), 184 (2022)
2. Bhambhvani, H.P., Zamora, A.: Deep learning enabled classification of mayo endoscopic subscore in patients with ulcerative colitis. Eur. J. Gastroenterol. Hepatol. **33**(5), 645–649 (2021)
3. Borgli, H., et al.: Hyperkvasir, a comprehensive multi-class image and video dataset for gastrointestinal endoscopy. Sci. Data **7**(1), 283 (2020)
4. Click, B., Ramos Rivers, C., Koutroubakis, I.E., Babichenko, D., Anderson, A.M., et al.: Demographic and clinical predictors of high healthcare use in patients with inflammatory bowel disease. Inflamm. Bowel Dis. **22**(6), 1442–1449 (2016)

5. Di Ruscio, M., et al.: Role of ulcerative colitis endoscopic index of severity (uceis) versus mayo endoscopic subscore (mes) in predicting patients response to biological therapy and the need for colectomy. Digestion **102**(4), 534–545 (2021)
6. Fan, Y., et al.: A novel deep learning-based computer-aided diagnosis system for predicting inflammatory activity in ulcerative colitis. Gastrointest. Endosc. (2022)
7. Gong, L., et al.: Automatic captioning of early gastric cancer using magnification endoscopy with narrow-band imaging. Gastrointest. Endosc. **96**(6), 929–942 (2022)
8. Ozawa, T., et al.: Novel computer-assisted diagnosis system for endoscopic disease activity in patients with ulcerative colitis. Gastrointest. Endosc. **89**(2), 416–421 (2019)
9. Pagnini, C., Di Paolo, M.C., Mariani, B.M., Urgesi, R., Pallotta, L., et al.: Mayo endoscopic score and ulcerative colitis endoscopic index are equally effective for endoscopic activity evaluation in ulcerative colitis patients in a real life setting. Gastroenterol. Insights **12**(2), 217–224 (2021)
10. Rabbenou, W., Ullman, T.A.: Risk of colon cancer and recommended surveillance strategies in patients with ulcerative colitis. Gastroenterol. Clin. **49**(4), 791–807 (2020)
11. Schroeder, K.W., Tremaine, W.J., Ilstrup, D.M.: Coated oral 5-aminosalicylic acid therapy for mildly to moderately active ulcerative colitis. New Engl. J. Med. **317**(26), 1625–1629 (1987)
12. Stidham, R.W., Liu, W., Bishu, S., Rice, M.D., Higgins, P.D., et al.: Performance of a deep learning model vs human reviewers in grading endoscopic disease severity of patients with ulcerative colitis. JAMA Netw. Open **2**(5), 1–10 (2019)
13. Sutton, R.T., Zaiane, O.R., Goebel, R., Baumgart, D.C.: Artificial intelligence enabled automated diagnosis and grading of ulcerative colitis endoscopy images. Sci. Rep. **12**(1), 1–10 (2022)
14. Taku, K., Britta, S., Chen, W.S., Ferrante, M., Shen, B., et al.: Ulcerative colitis (primer). Nat. Rev. Dis. Primers **6**(1) (2020)
15. Travis, S.P., Schnell, D., Krzeski, P., Abreu, M.T., Altman, D.G., et al.: Developing an instrument to assess the endoscopic severity of ulcerative colitis: the ulcerative colitis endoscopic index of severity (uceis). Gut **61**(4), 535–542 (2012)
16. Ungaro, R., Mehandru, S., Allen, P., Peyrin-Biroulet, L., Colombel, J.: Colitis ulcerosa. Lancet **389**(10080), 1756–1770 (2017)
17. Vashist, N.M., et al.: Endoscopic scoring indices for evaluation of disease activity in ulcerative colitis. Cochrane Datab. Syst. Rev. (1) (2018)
18. Xu, Z., Ali, S., East, J., Rittscher, J.: Additive angular margin loss and model scaling network for optimised colitis scoring. In: 2022 IEEE 19th International Symposium on Biomedical Imaging (ISBI), pp. 1–5. IEEE (2022)
19. Xu, Z., Ali, S., Gupta, S., Leedham, S., East, J.E., Rittscher, J.: Patch-level instance-group discrimination with pretext-invariant learning for colitis scoring. In: Machine Learning in Medical Imaging, pp. 101–110 (2022)

Detection and Segmentation

Multispectral 3D Masked Autoencoders for Anomaly Detection in Non-Contrast Enhanced Breast MRI

Daniel M. Lang[1,2(✉)], Eli Schwartz[3,4], Cosmin I. Bercea[1,2,5], Raja Giryes[4], and Julia A. Schnabel[1,2,5,6]

[1] Helmholtz Munich, Neuherberg, Germany
[2] Technical University of Munich, Garching, Germany
daniel.lang@tum.de
[3] IBM Research AI, Tel Aviv, Israel
[4] Tel Aviv University, Tel Aviv, Israel
[5] Helmholtz AI, Munich, Germany
[6] King's College London, London, UK

Abstract. Mammography is commonly used as an imaging technique in breast cancer screening but comes with the disadvantage of a high overdiagnosis rate and low sensitivity in dense tissue. Dynamic contrast enhanced (DCE)-magnetic resonance imaging (MRI) features higher sensitivity but requires time consuming dynamic imaging and injection of contrast media, limiting the capability of the technique as a widespread screening method. In this work, we extend the masked autoencoder (MAE) approach to perform anomaly detection on volumetric, multispectral MRI. This new model, coined masked autoencoder for medical imaging (MAEMI), is trained on two non-contrast enhanced breast MRI sequences, aiming at lesion detection without the need for intravenous injection of contrast media and temporal image acquisition, paving the way for more widespread use of MRI in breast cancer diagnosis. During training, only non-cancerous images are presented to the model, with the purpose of localizing anomalous tumor regions during test time. We use a public dataset for model development. Performance of the architecture is evaluated in reference to subtraction images created from DCE-MRI. Code has been made publicly available: https://github.com/LangDaniel/MAEMI.

1 Introduction

Breast cancer depicts the most commonly diagnosed cancer and leading cancer-related cause of death in women worldwide [15]. In general, early cancer detection allows for effective treatment and improves survival substantially [7]. Mammography is most frequently employed as a breast cancer screening technique and was shown to be able to reduce mortality by up to 30% [8]. However, sensitivity

Supplementary Information The online version contains supplementary material available at https://doi.org/10.1007/978-3-031-45350-2_5.

© The Author(s), under exclusive license to Springer Nature Switzerland AG 2023
S. Ali et al. (Eds.): CaPTion 2023, LNCS 14295, pp. 55–67, 2023.
https://doi.org/10.1007/978-3-031-45350-2_5

of the approach depends on breast density. For extremely dense tissue 40% of cancers are not detected [27], while in other cases an overdiagnosis rate of 31% has been reported [17].

DCE-MRI refers to the acquisition of images before, during and after intravenous injection of contrast media, which improves the signal intensity of neoangiogenically induced vascular changes that allows for better detection of lesions [26]. The technique features a higher sensitivity for detection of breast cancer, when compared to mammography [23]. However, dynamic imaging and long scan times in combination with high costs limit widespread use for screening [16]. Moreover, injection of gadolinium-based contrast agents were identified to be able to cause side effects, like nephrogenic systemic fibrosis, which raised concerns about the broader health impact of the technique [10]. Additionally, patient motion during imaging can lead to artifacts in subtraction images, resulting in incorrect identification of tumor lesions.

Deep learning algorithms may have the power to overcome the need for contrast injection and temporal imaging and could be trained on *non-contrast enhanced* MRI for tumor detection. However, networks require large amounts of data with detailed voxel level annotations from medical experts, when trained in a supervised fashion.

Unsupervised anomaly detection aims to identify abnormal cases without the requirement for labeled examples. This can be achieved by reconstruction-based methods. Models are trained to recover their input, while restrictions on the architecture are applied. Such restrictions can be imposed by information bottlenecks [5] or by the alteration of input images via application of noise [14] or removal of image parts [29]. During training, normal examples are shown to the model. In this way, the model is only able to reconstruct image parts stemming from the normal distribution reasonably well, while abnormal image parts result in higher error rates that can be utilized to generate anomaly maps during test time. Applications of anomaly detection (AD) in the medical imaging domain include but are not limited to: chest X-ray, optical coherence tomography (OCT) and mammography. Most research in the field has been conducted on MRI of the brain, e.g. [2,14].

We demonstrate the capability of an unsupervised model for anomaly detection on non-contrast enhanced breast MRI. Namely, the MAE approach of *He et al.* [12] is here extended to be able to handle 3D multispectral medical imaging data, and trained for anomaly detection on a public breast cancer dataset. Performance of the model is evaluated in reference to subtraction images generated from DCE-MRI.

1.1 Contribution

In this work we demonstrate the capability of anomaly detection models for identification of tumor lesion on non-contrast enhanced MRI. To do so, we remodel the MAE architecture and extend and further develop the approach of *Schwartz et al.* [24], enabling unsupervised anomaly detection on 3D multispectral medical imaging data. Definition of input patches of the vision transformer (ViT) architecture is advanced and the positional embedding employed by *He et al.* [12] is refined. During training only healthy, non-cancerous breast MRIs are shown to the model, aiming to identify breast lesions as anomalies during test time. To the best of our knowledge, we are the first to make the following contributions:

- We extend and further adapt MAEs to perform anomaly detection on 3D multispectral medical imaging.
- We demonstrate the capability of unsupervised anomaly detection to identify pathologies in non-contrast enhanced breast MRI.
- We show that performance of anomaly detection algorithms on non-contrast enhanced MRI is on par with that of DCE-MRI generated subtraction images, paving the way for a more widespread use of MRI in breast cancer diagnosis.

2 Related Work

Lesions detection on breast MRI has been studied by training of a deep Q-network [18], modification of RetinaNet [1], and utilization of the ResNet architecture [13]. Synthetic contrast enhancement was investigated by training of a generative adversarial network (GAN) [19], utilization of a U-Net like architecture [4], and combination of a encoder-decoder architecture with a LSTM layer [11].

Notably, all of those approaches were trained on DCE-MRI data, relying on injection of contrast media, and employed supervised training. To the best of our knowledge, we are the first to perform unsupervised anomaly detection on non-contrast enhanced MRI, without the need for contrasted enhanced images, even during training.

The ability of deep convolutional autoencoders (AEs) to perform reconstruction based unsupervised anomaly detection on imaging data has been investigated by several studies. *Kascenas et al.* [14] trained a denoising AE on brain MRI, such that unhealthy pathologies were removed during test time. *Zavrtanik et al.* [29] developed a convolutional AE, masking part of the input to perform inpainting on natural imaging and video data.

He et al. [12] developed the masked autoencoder (MAE) model on 2D natural images, to solve a classification downstream task. *Schwartz et al.* [24] modified the architecture to allow for anomaly detection on natural imaging data. Here, we advance the approach to be able to perform anomaly detection on 3D medical imaging data stemming from different input sequences.

Prabhakar et al. [20] improved the initial MAE architecture by incorporation of a contrastive and an auxiliary loss term, and trained their architecture on classification of brain MRI. Our model relies on image reconstruction and computation of a voxel-wise difference in the downstream task. Therefore, we use unaltered mean squared error (MSE) as a loss.

Xu et al. [28] developed SwinMAE, embedding Swin transformer in the MAE architecture, for supervised segmentation of parotid tumors in the downstream task. They trained their model on MRI, feeding slices from different sequences to the color channels of a 2D model. In contrast, we train our model in 4D, handling different 3D MRI sequences fully unsupervised.

Due to its high masking ratio, anomalies are likely to be removed by MAE. This led *Tian et al.* [25] to employ the model on anomaly detection in 2D colonoscopy and X-ray data. They introduced *memory-augmented self-attention* and a *multi-level cross-attention operator* in the underlying ViT architecture, to limit dependency on random masking. In contrast, we train our model on multispectral 3D data, without any modifications applied to the ViT architecture, maximizing the likelihood for anomalies to be removed by application of a high number of random masks.

3 Dataset

We used the public Duke-Breast-Cancer-MRI cohort [21,22] from The Cancer Imaging Archive [6]. The set includes axial breast MRI data of 922 patients comprising a non-fat saturated pre-contrast T1-weighted sequence, a fat-saturated T1-weighted pre-contrast sequence and several post-contrast fat-saturated T1-weighted sequences. Tumor lesion annotations were given in the form of bounding boxes and segmentation maps, labeling breast tissue, were provided. We standardized all images to the same voxel spacing of $0.75mm \times 0.75mm \times 1.0mm$ and cropped them to involve the chest area only. Intensity values of the images were normalized to a mean and standard deviation value of 0.5 and 0.25 per image. Cases involving bilateral breast cancer were removed from the cohort. Thus, each of the remaining patients exhibited one breast that contained a tumor lesion, treated as *abnormal/unhealthy*, and another breast not affected by cancer, treated as *normal/healthy*. The dataset was split into a training set of 745 patients, a validation set of 50 patients and a test set of 100 patients. Notably, no *normal - abnormal* pairs of the same patient were involved in different datasets. Due to the high memory requirements of transformer models, MRI-patches of size $240 \times 168 \times 8$ voxels in lateral-posterior-superior (LPS) directions were cropped from both MRI sequences to be then divided into ViT-patches and processed by the encoder.

Following clinical routine, subtraction images S between the image acquired before I_{pre} and all of the m images acquired after injection of contrast media I^k_{post} were computed:

$$S = \frac{1}{m}\sum_{k=0}^{m}\left(I_{\text{pre}} - I^k_{\text{post}}\right)^2 * \min\nolimits_{3\times3\times2}, \tag{1}$$

with a minimum filter of size $3 \times 3 \times 2$ applied for noise removal.

4 Method

A scheme of our model can be seen in the lower part of Fig. 1. We modified the ViT architecture of *Dosovitskiy et al.* [9] to 3D multispectral MRI, i.e. the positional embedding and definition of ViT-patches was redefined and enhanced to incorporate a third dimension, see Fig. 2.

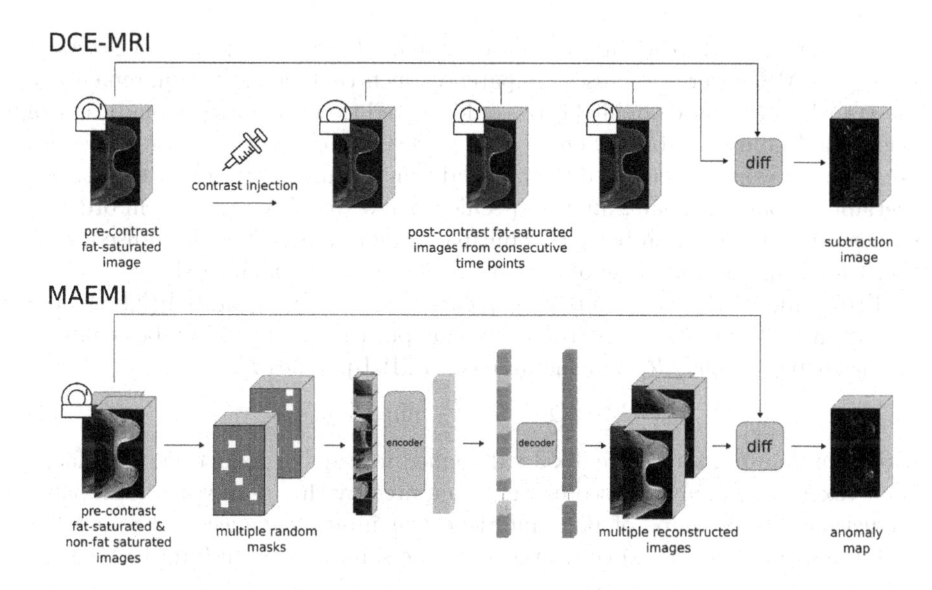

Fig. 1. DCE-MRI imaging vs. MAEMI. For DCE-MRI, several MRIs before, during and after injection of contrast media are acquired. MAEMI uses different random masks for generation of pseudo-healthy recovered images. Both methods construct error maps by calculation of the mean squared error difference between each of the post-contrast/reconstructed images with the pre-contrast/ uncorrupted image.

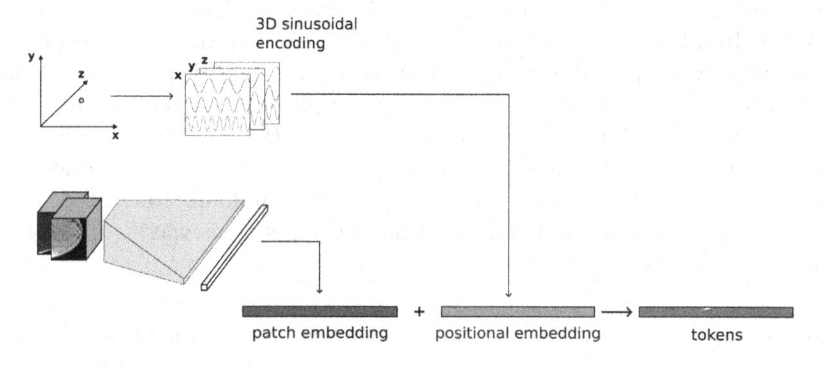

Fig. 2. 3D multispectral embedding. ViT-patches from different sequences are combined and embedded using a 3D convolutional layer. The positional embedding is advanced to 3D, adding a additional dimension to the sinusoidal encoding. As for the MAE model [12], no mask tokens are used in the encoder.

This 3D-ViT was then embedded in the MAE approach [12], which had to be extended to be able to handle three dimensional input from different sequences. To do so, generation of mask tokens and random masking of ViT-patches had to be remodeled. The approach, coined masked autoencoder for medical imaging (MAEMI), uses a 3D-ViT with 12 transformer blocks and a embedding dimension of 768 as encoder, while for the decoder a embedding dimension of 384 and a depth of 4 has been chosen.

In addition, we further improved the anomaly detection approach of *Schwartz et al.* [24]. MRIs were processed by patches, such that memory requirements of the transformer based architecture could be reduced, enabling exploitation of whole MRI volumes from different sequences. Only examples of non-cancerous breast tissue were presented to the model during training. At inference, a strided overlapping patch scheme, further specified below, has been chosen, in order to reduce artifacts on patch borders. Multiple random masks have been used, such that the likelihood for the anomaly to be removed was maximized.

Error maps E_i^s per input sequence s, non-fat saturated (NFS) and fat saturated (FS), were constructed by computation of the MSE between the reconstructed patches R_i^{s} and the unmasked MRI-patches I^{s}:

$$E_i^{\mathrm{s}} = (I^{\mathrm{s}} - R_i^{\mathrm{s}})^2 * \min\nolimits_{3\times3\times2}, \tag{2}$$

with a minimum filter of size $3 \times 3 \times 2$ applied for further reduction of artifacts. On a voxel level, final error scores were computed by the mean value of all patch predictions. Error maps of both multispectral input sequences, NFS and FS, were then summed up and convolved with the same minimum filter as before:

$$E = \frac{1}{2}\left(E^{\mathrm{NFS}} + E^{\mathrm{FS}}\right) * \min\nolimits_{3\times3\times2}, \tag{3}$$

for generation of a final image level anomaly map.

Training Specifics. During training, only MRI-patches of healthy breasts, containing no tumor lesions, were shown to the model with patches being cropped randomly. In addition to random cropping, random flipping on the coronal and sagittal plane was applied as a augmentation technique during training. A batch size of 6 and a learning rate of 10^{-3} were applied. Each model was trained for 1000 epochs. Weights of the trained model of *He et al.* [12], developed on ImageNet, were used to initialize the transformer layers in the encoder, while weights of the encoding layer and the decoder were randomly initialized. During test time, a stride of $64 \times 42 \times 2$ performing 6 repetitions was used to process whole MRI volumes.

Evaluation. We used voxel wise area under the receiver operating characteristics curve and average precision as performance measures. Only voxels inside the breast tissue segmentation masks were taken into account for computation, as injection of contrast media also leads to an uptake in tissue outside of the breast area, see Fig. 7 (Supplementary Material). Furthermore, incorporation of voxels outside of patients' body, containing air and therefore obviously no tumor tissue, would lead to an overestimation of performance.

Utilizing average precision (AP) as a performance measure, one has to consider the large imbalance between normal and abnormal tissue labels, leading to an expected small baseline performance. This baseline can be computed by the number of voxels inside the bounding box normalized by the number of voxels inside the breast tissue segmentation mask and is given by a value of 0.046 for the test set.

5 Results

ViT-patch size and masking ratio have been varied for hyperparameter tuning. The best performing model featured a masking ratio of 90% and a ViT-patch size of $8 \times 8 \times 2$. area under the receiver operating characteristics curve (AUROC) and AP results are shown in Table 1. Example results are shown in Fig. 3.

Table 1. MAEMI achieves a higher AUROC, while DCE-MRI features a higher AP.

	AUROC	AP
MAEMI	**0.732**	0.081
DCE-MRI	0.705	**0.127**

Table 2. Ablation study on different (axial) ViT-patch sizes, for a fixed masking ratio of 90%. Smaller patch sizes lead to higher performance.

ViT-patch size	AUROC	AP
$6 \times 6 \times 4$	0.724	0.0784
$8 \times 8 \times 4$	0.712	0.0777
$12 \times 12 \times 4$	0.660	0.0750
$24 \times 24 \times 4$	0.546	0.0560

Ablation Studies. We studied the influence of the patch size and masking ratio on model performance. Figure 4 presents the dependency of AUROC and AP on the masking ratio for a fixed ViT-patch size of $8 \times 8 \times 2$. Example results are given in Fig. 6 in the Supplementary Material. Dependency on the ViT-patch size for a fixed masking ratio of 90% is given in Table 2. The Nvidia RTX A6000 used for training, featuring 48 GB of memory, only allowed for a smallest size of $8 \times 8 \times 2$, i.e. a size of $6 \times 6 \times 2$ could not be probed. Therefore, the slice dimension of ViT-patches was fixed at a value of 4 pixels, probing only different axial sizes.

6 Discussion

We developed a multispectral 3D transformer-based anomaly detection model, that has been trained to identify breast lesions on non-contrast enhanced MRI. Model performance was at the same level as for DCE-MRI generated subtraction images. To the best of our knowledge, we are the first to demonstrate the ability of breast lesion identification relying solely on non-contrast enhanced MRI.

Eliminating the need for DCE would result in a drastic reduction of costs and acquisition time, two major factors limiting widespread application of MRI in breast cancer screening [16]. Patients would be spared from intravenous injection of contrast media, which is known to be able to cause side effects, and artifacts in subtraction images, introduced by patient motion, can be removed as a source of error. Therefore, we paved the way for a more widespread application of MRI in breast cancer diagnosis. Clinical differences and benefits between MAEMI and DCE-MRI will still need to be investigated in a larger clinical study.

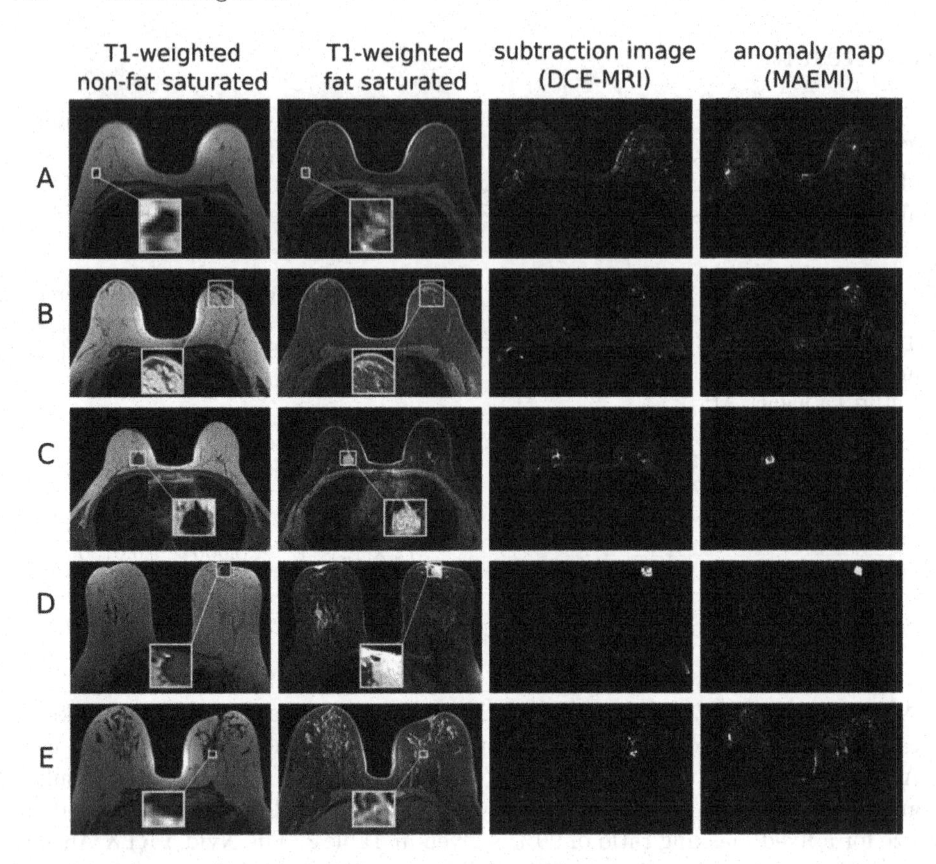

Fig. 3. Example results. The first two columns show the non-contrast enhanced images used as an input to the anomaly detection model, and the last two columns present subtraction images generated by DCE-MRI and anomaly detection maps generated by MAEMI, respectively. For patients in rows A, B and C, anomaly maps show superior performance over subtraction images. For patient D, both methods are able to identify the pathology. For patient E, our model only detects the borders of the pathology, while the subtraction image identifies the lesion.

MAEMI achieved a higher AUROC while DCE-MRI generated subtraction images resulted in better AP performance. Ground truth annotations were given in the form of bounding boxes, depicting only a rough delineation of tumor tissue. Determination of an optimal performance measure remains an active area of research [3].

We found an optimal masking ratio of 90% for our model. *Tian et al.* [25] employed the standard masking ratio of 75% for their 2D model, and did not report any ablation studies. Higher masking ratios increase the likelihood of the anomaly to be removed, leading better results. However, for very high ratios the model is not able to recover the input reasonably well, see Fig. 6. This is likely to causes the steep decline in Fig. 4.

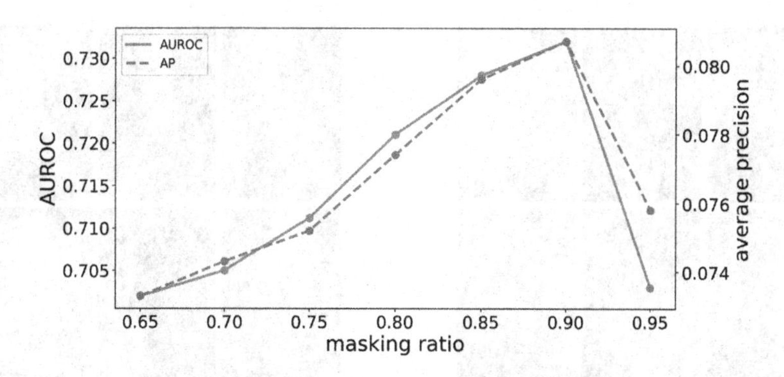

Fig. 4. Ablation study on the masking ratio for a fixed ViT-patch size of $8 \times 8 \times 2$. High masking ratios lead to better performance, with an optimum reached at 90%. Afterwards, performance suffers from a steep decline. This behaviour is likely to be caused by the fact that larger masking ratios increase the probability of the anomaly to be removed. However, at a certain level the model is not able to recover the input images, causing the steep decline.

In future work, we will test the performance of other anomaly detection methods for identification of tumor lesion on non-contrast enhanced MRI. However, models are most often developed on 2D data, i.e. on slices of brain MRI, and have to be advanced to be able to handle 3D data first. Furthermore, the capability of anomaly detection to replace injection of contrast media will also be studied for other entities, e.g. in liver MRI.

Data Use Declaration. All data used for this study is publicly available from The Cancer Imaging Archive [6,21] under the CC BY-NC 4.0 license. Code has been published at: https://github.com/LangDaniel/MAEMI.

Acknowledgements. DML was in part financed by the Helmholtz Information and Data Science Academy (HIDA) under the “Israel Exchange Program”.

Supplementary Material

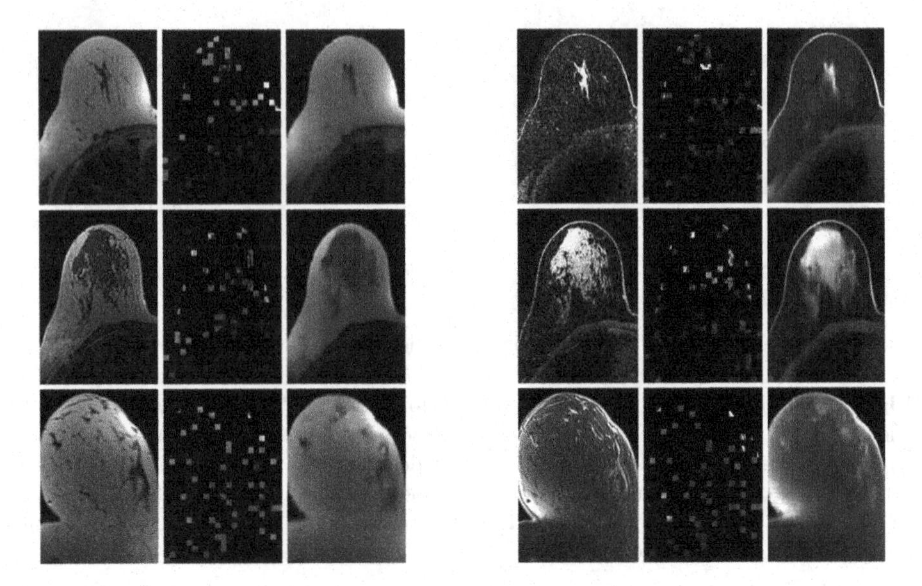

Fig. 5. Single reconstruction examples. The left block shows axial slices of T1 non-fat saturated MRI-patches and the right block T1 fat saturated slices. The first column shows unaltered MRI-patches, the second column the masked model input and the third column the MRI-patches recovered by MAEMI. Examples represent a masking ratio of 90% (for the whole 3D patch) and a ViT-patch size of $8 \times 8 \times 2$. During inference, multiple masks are generated to maximize likelihood of anomalies to be removed.

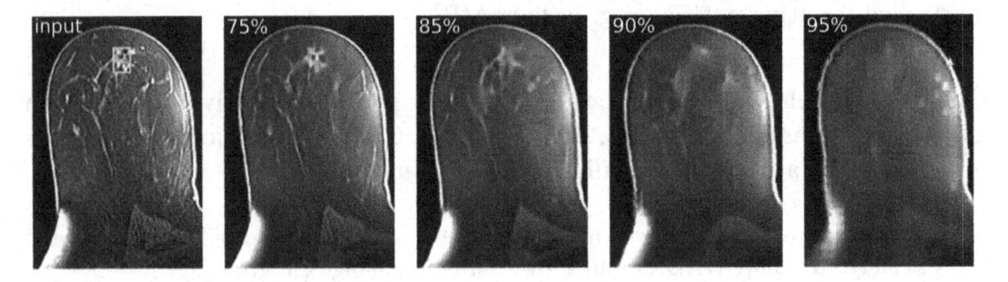

Fig. 6. Reconstruction examples for different masking ratios and a fixed ViT-patch size of $8 \times 8 \times 2$. Very high ratios (>90%) lead blurry images, while moderate ratios allow for reconstruction of detailed structures. On the one hand, reconstructions need to be detailed enough to reduce the amount of false positive findings. On the other, high masking ratios increase the likelihood of the anomaly to be removed, leading to a better true positive rate.

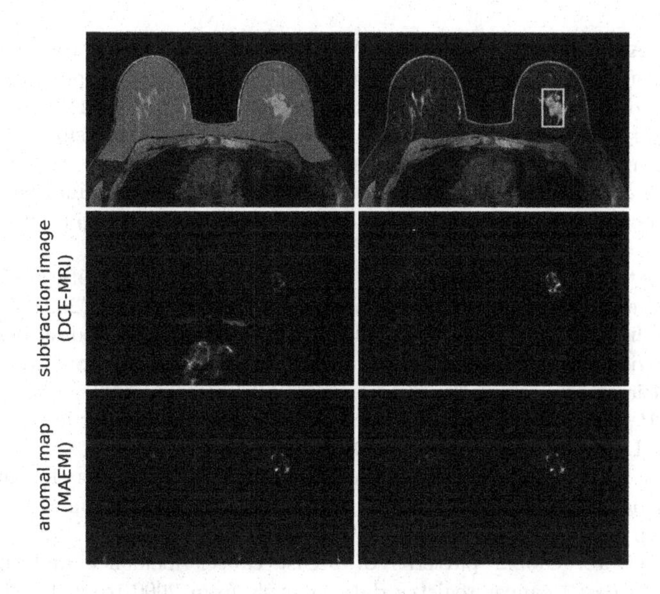

Fig. 7. Subtraction images and anomaly maps were multiplied with segmentation masks to remove anomalies lying obviously outside of the breast tissue. This is mainly needed as contrast agent is also taken up in organs outside of the breast. The left column shows the raw subtraction/anomaly map, and the right column the raw maps multiplied with the segmentation mask of the image in the upper left corner. Performance metrics were only calculated for voxels lying inside the segmentation mask, limiting the influence of trivial predictions, i.e. voxels that represent air do not containing any anomalies.

References

1. Ayatollahi, F., Shokouhi, S.B., Mann, R.M., Teuwen, J.: Automatic breast lesion detection in ultrafast DCE-MRI using deep learning. Med. Phys. **48**(10), 5897–5907 (2021)
2. Baur, C., Denner, S., Wiestler, B., Navab, N., Albarqouni, S.: Autoencoders for unsupervised anomaly segmentation in brain MR images: a comparative study. Med. Image Anal. **69**, 101952 (2021)
3. Bercea, C.I., Wiestler, B., Rueckert, D., Schnabel, J.A.: Generalizing unsupervised anomaly detection: towards unbiased pathology screening. In: Medical Imaging with Deep Learning (2023)
4. Bône, A., et al.: Contrast-enhanced brain MRI synthesis with deep learning: key input modalities and asymptotic performance. In: 2021 IEEE 18th International Symposium on Biomedical Imaging (ISBI), pp. 1159–1163. IEEE (2021)
5. Chow, J.K., Su, Z., Wu, J., Tan, P.S., Mao, X., Wang, Y.H.: Anomaly detection of defects on concrete structures with the convolutional autoencoder. Adv. Eng. Inf. **45**, 101105 (2020)
6. Clark, K., et al.: The Cancer Imaging Archive (TCIA): maintaining and operating a public information repository. J. Dig. Imaging **26**, 1045–1057 (2013)
7. Crosby, D., et al.: Early detection of cancer. Science **375**(6586), eaay9040 (2022)

8. Dibden, A., Offman, J., Duffy, S.W., Gabe, R.: Worldwide review and meta-analysis of cohort studies measuring the effect of mammography screening programmes on incidence-based breast cancer mortality. Cancers **12**(4), 976 (2020)
9. Dosovitskiy, A., et al.: An image is worth 16×16 words: transformers for image recognition at scale. arXiv preprint arXiv:2010.11929 (2020)
10. Fraum, T.J., Ludwig, D.R., Bashir, M.R., Fowler, K.J.: Gadolinium-based contrast agents: a comprehensive risk assessment. J. Magn. Reson. Imaging **46**(2), 338–353 (2017)
11. Han, L., et al.: Synthesis-based imaging-differentiation representation learning for multi-sequence 3D/4D MRI. arXiv preprint arXiv:2302.00517 (2023)
12. He, K., Chen, X., Xie, S., Li, Y., Dollár, P., Girshick, R.: Masked autoencoders are scalable vision learners. In: Proceedings of the IEEE/CVF Conference on Computer Vision and Pattern Recognition, pp. 16000–16009 (2022)
13. Herent, P., et al.: Detection and characterization of MRI breast lesions using deep learning. Diagn. Interv. Imaging **100**(4), 219–225 (2019)
14. Kascenas, A., Pugeault, N., O'Neil, A.Q.: Denoising autoencoders for unsupervised anomaly detection in brain MRI. In: International Conference on Medical Imaging with Deep Learning, pp. 653–664. PMLR (2022)
15. Lei, S., et al.: Global patterns of breast cancer incidence and mortality: a population-based cancer registry data analysis from 2000 to 2020. Cancer Commun. **41**(11), 1183–1194 (2021)
16. Leithner, D., Moy, L., Morris, E.A., Marino, M.A., Helbich, T.H., Pinker, K.: Abbreviated MRI of the breast: does it provide value? J. Magn. Reson. Imaging **49**(7), e85–e100 (2019)
17. Løberg, M., Lousdal, M.L., Bretthauer, M., Kalager, M.: Benefits and harms of mammography screening. Breast Cancer Res. **17**(1), 1–12 (2015)
18. Maicas, G., Carneiro, G., Bradley, A.P., Nascimento, J.C., Reid, I.: Deep reinforcement learning for active breast lesion detection from DCE-MRI. In: Descoteaux, M., et al. (eds.) MICCAI 2017. LNCS, vol. 10435, pp. 665–673. Springer, Cham (2017). https://doi.org/10.1007/978-3-319-66179-7_76
19. Müller-Franzes, G., et al.: Using machine learning to reduce the need for contrast agents in breast MRI through synthetic images. Radiology **307**(3), e222211 (2023)
20. Prabhakar, C., Li, H.B., Yang, J., Shit, S., Wiestler, B., Menze, B.: ViT-AE++: improving vision transformer autoencoder for self-supervised medical image representations. arXiv preprint arXiv:2301.07382 (2023)
21. Saha, A., et al.: Dynamic contrast-enhanced magnetic resonance images of breast cancer patients with tumor locations. Cancer Imaging Arch. (2021)
22. Saha, A., et al.: A machine learning approach to radiogenomics of breast cancer: a study of 922 subjects and 529 DCE-MRI features. Brit. J. cancer **119**(4), 508–516 (2018)
23. Sardanelli, F., et al.: Sensitivity of MRI versus mammography for detecting foci of multifocal, multicentric breast cancer in fatty and dense breasts using the whole-breast pathologic examination as a gold standard. Am. J. Roentgenol. (2012)
24. Schwartz, E., et al.: MAEDAY: MAE for few and zero shot AnomalY-Detection. arXiv preprint arXiv:2211.14307 (2022)
25. Tian, Y., et al.: Unsupervised anomaly detection in medical images with a memory-augmented multi-level cross-attentional masked autoencoder. arXiv preprint arXiv:2203.11725 (2022)
26. Turnbull, L.W.: Dynamic contrast-enhanced MRI in the diagnosis and management of breast cancer. NMR Biomed. Int. J. Dev. Dev. Appl. Magn. Reson. Vivo **22**(1), 28–39 (2009)

27. Wanders, J.O., et al.: Volumetric breast density affects performance of digital screening mammography. Breast Cancer Res. Treat. **162**, 95–103 (2017)
28. Xu, Z., et al.: Swin MAE: masked autoencoders for small datasets. arXiv preprint arXiv:2212.13805 (2022)
29. Zavrtanik, V., Kristan, M., Skočaj, D.: Reconstruction by inpainting for visual anomaly detection. Pattern Recogn. **112**, 107706 (2021)

Non-redundant Combination of Hand-Crafted and Deep Learning Radiomics: Application to the Early Detection of Pancreatic Cancer

Rebeca Vétil[1,2](✉), Clément Abi-Nader[2], Alexandre Bône[2], Marie-Pierre Vullierme[3], Marc-Michel Rohé[2], Pietro Gori[1], and Isabelle Bloch[1,4]

[1] LTCI, Télécom Paris, Institut Polytechnique de Paris, Paris, France
[2] Guerbet Research, Villepinte, France
rebeca.vetil@guerbet.com
[3] Department of Radiology, Hospital of Annecy-Genevois, Université de Paris, Paris, France
[4] Sorbonne Université, CNRS, LIP6, Paris, France

Abstract. We address the problem of learning Deep Learning Radiomics (DLR) that are not redundant with Hand-Crafted Radiomics (HCR). To do so, we extract DLR features using a VAE while enforcing their independence with HCR features by minimizing their mutual information. The resulting DLR features can be combined with hand-crafted ones and leveraged by a classifier to predict early markers of cancer. We illustrate our method on four early markers of pancreatic cancer and validate it on a large independent test set. Our results highlight the value of combining non-redundant DLR and HCR features, as evidenced by an improvement in the Area Under the Curve compared to baseline methods that do not address redundancy or solely rely on HCR features.

Keywords: Early Diagnosis · Pancreatic Cancer · Radiomics · Variational Autoencoders · Mutual Information

1 Introduction

Computational methods in medical imaging hold the potential to support radiologists in the early diagnosis of cancer, either by detecting small-size abnormal neoplasms [14], or even earlier in the disease course by recognizing indirect signs of malignancy. Such signs are usually subtle and organ-dependent, thus requiring a time-consuming and demanding clinical assessment. For example, in the case of pancreatic cancer, radiologists analyze the overall shape of the organ, check for fat replacement and note whether the pancreas shows atrophy and/or senile characteristics [7,18,19]. The identification of cancerous signs using automated tools can be based on radiomics, which are descriptors of texture and

© The Author(s), under exclusive license to Springer Nature Switzerland AG 2023
S. Ali et al. (Eds.): CaPTion 2023, LNCS 14295, pp. 68–82, 2023.
https://doi.org/10.1007/978-3-031-45350-2_6

shape of a medical image, computed based on spatial relationships between voxels and their intensity distribution [11,12]. Radiomics can be divided into two categories: (i) Hand-Crafted Radiomics (HCR), which are based on predefined mathematical formulas [11,12]; (ii) Deep Learning Radiomics (DLR), estimated using deep neural networks [10,23], which may unveil additional complex relationships between voxels. HCR are generally extracted by open-source frameworks such as pyradiomics [24]. While such tools facilitate the standardization of the HCR, they only provide a limited number of predefined features. On the other hand, DLR features are typically extracted using either discriminative or generative models. Discriminative models frequently rely on one or multiple simple CNNs [3–5,13,20]. To prevent overfitting, some methods extract DLR by utilizing pretrained models trained on large datasets like ImageNet [3,5,20]. The deep neural networks commonly employed for computing these DLR features consist of multiple layers, with each layer producing potential features as its output. As a result, the choice of the layers to retain varies, with each method employing different heuristics to identify them [5,20]. In the realm of generative models, auto-encoder (AE) networks are widely used [2]. AEs encode an image in a latent vector that is subsequently used to reconstruct the original image. This latent vector is considered to encapsulate the most descriptive features of the input image, making it a natural choice for representing the DLR [10,21].

The two types of radiomics are complementary: the computation of DLR is data-driven, which ensures that the extracted features are adapted to a specific problem or type of data. On the other hand, the predefined and generic definitions of HCR may make them less adapted for a given specific task, but favors generalization and interpretability. Therefore, it has been recently proposed to combine HCR with DLR, arguing that this approach could result in an improved feature set for predictive or prognostic models [2]. The literature reports two main approaches to perform this combination: decision-level methods that train separate classifiers on DLR and HCR before aggregating their predictions [3,5,16], and feature-level methods that concatenate the two types of radiomics in a single feature vector which is then leveraged by a classifier [4,13,20]. These approaches extract HCR and DLR features independently, without guaranteeing complementarity between the two sets of features. As a result, the extracted DLR may be highly redundant with the HCR, limiting the value of their combination.

Given this context, we propose to extract DLR features that will complement the information already contained in the HCR. Our contributions are two-fold:

- A deep learning method, based on the VAE framework [9], that extracts non-redundant DLR features with respect to a predetermined set of HCR. This is achieved by minimizing the mutual information between the two types of radiomics during the training of the VAE. The resulting HCR and DLR features are leveraged to predict early markers of cancer.
- Validation of the proposed approach in the case of pancreatic cancer, using 2319 training and 1094 test subjects collected from 9 medical institutions with a split performed at the institution level. This is all the more important as most combination approaches have been solely evaluated in a cross-validation setting on mono-centric data [3,5,16].

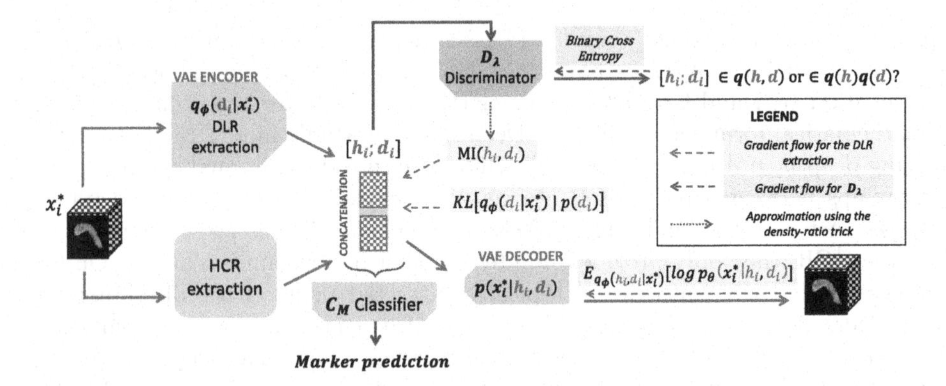

Fig. 1. Overview of our method. Starting from a masked image, Hand-Crafted Radiomics (HCR) are calculated analytically, while Deep Learning Radiomics (DLR) are extracted by the encoder of a VAE. These two types of radiomics are subsequently combined and given to the decoder for image reconstruction. The independence of HCR and DLR is enforced by the minimization of the Mutual Information (MI). The latter is approximated by the density-ratio trick [8], involving a discriminator $\mathcal{D}_\lambda$. Following the training of the VAE, a classifier $\mathcal{C}_M$ can be trained using both the HCR and DLR features to predict a specific marker of interest.

2 Method

Our method, illustrated in Fig. 1, relies on a generative model that recreates a 3D input image from the concatenation of HCR and DLR features. Feature extraction is done analytically for the HCR and through a VAE encoder for the DLR. Independence between the features is encouraged through the minimization of their mutual information, which is estimated by a discriminator relying on the density-ratio trick [8]. Finally, the resulting features are given to a classifier for cancer marker prediction.

Generative Framework. Let $x \in \mathbb{R}^V$ be a 3D image acquired via a standard imaging technique, and $y \in \{0,1\}^V$ the corresponding binary segmentation mask of a given organ, with V the number of voxels. In order to focus on a specific organ and facilitate the extraction of specific features, we work on the masked image $x^* = x \times y$. We postulate the existence of a generative model enabling us to create an image x^* from a low-dimensional representation space $[h, d]$ where $h \in \mathrm{R}^{N_h}$ and $d \in \mathrm{R}^{N_d}$ represent the HCR and DLR features with N_h and N_d being the number of hand-crafted and deep features, respectively. Assuming that x^* follows an independent and identically distributed Gaussian distribution, and that f_θ is a non-linear function mapping the concatenation of vectors $[h, d]$ to the masked image x^*, we hypothesize the following generative process:

$$p_\theta(x^* \mid y, h, d) = \prod_{v=1/y_v=1}^{V} \frac{1}{\sqrt{2\pi\sigma^2}} \exp \frac{(x_v^* - f_\theta([h, d])_v)^2}{2\sigma^2} \quad (1)$$

HCR and DLR Features Computation. We place ourselves within the VAE framework [9] and assume that $p(d)$ follows a Gaussian distribution with zero mean and identity covariance. HCR features are calculated analytically, while DLR features are computed by introducing the approximate posterior distribution $q_\phi(d \mid x^*)$. We hypothesize $q_\phi(d \mid x^*) \sim \mathcal{N}(\mu_\phi(x^*), \sigma^2_\phi(x^*)\mathbf{I})$, and maximize a lower bound of the marginal log-likelihood $\log p_\theta(x^* \mid y)$. We obtain the following loss function:

$$\mathcal{L}_{\text{VAE}} = -\mathbb{E}_{q_\phi(d|x^*)}[\log(p_\theta(x^* \mid y, h, d))] + KL[q_\phi(d \mid x^*) \mid p(d)] \tag{2}$$

where KL refers to the Kullback-Leibler divergence.

Mutual Information Minimization. To promote the independence between HCR and DLR features, we propose to minimize their Mutual Information (MI), expressed here as $KL[q(h,d) \mid q(h)q(d)]$, where $q(h,d)$ represents the joint distribution of the DLR and HCR features, and $q(h)q(d)$ the product of their marginal distributions. These terms involve mixtures with a large number of components, making them intractable. Moreover, obtaining the direct Monte Carlo estimate necessitates processing the entire dataset in a single pass. Thus, we sample from these distributions to compute the MI: to sample from $q(h,d)$, we randomly choose an image x_i^*, extract its HCR features h_i as well as its DLR features d_i using the VAE encoder, and concatenate them. Samples from $q(h)q(d)$ are obtained by concatenating vectors h_k and d_j with $k \neq j$. Finally, to compute the MI, we need to compute the density-ratio between $q(h,d)$ and $q(h)q(d)$. To do so, we resort to the density-ratio trick [8], which consists in introducing a discriminator $\mathcal{D}_\lambda([h,d])$ able to discriminate between samples from $q(h,d)$ and samples from $q(h)q(d)$. Thus, we obtain:

$$KL[q(h,d) \mid q(h)q(d)] = \mathbb{E}_{q(h,d)}\left[\log \frac{q(h,d)}{q(h)q(d)}\right] \approx \sum_i \text{ReLU}\left(\left[\log \frac{\mathcal{D}_\lambda(h_i,d_i)}{1-\mathcal{D}_\lambda(h_i,d_i)}\right]\right). \tag{3}$$

where the ReLU function forces the estimate of the MI to be positive, which prevents from back-propagating wrong estimates of the density-ratio. $\mathcal{D}_\lambda$ implementation is detailed in Sect. A.1 of the appendix.

Optimization. The final loss function is:

$$\mathcal{L} = \mathcal{L}_{\text{VAE}} + \kappa KL[q(h,d) \mid q(h)q(d)] \tag{4}$$

This loss function is composed of two terms: the left-hand term, which is the common VAE loss function and promotes the reconstruction of the masked image while regularizing the approximate posterior distribution; and the right-hand term which minimizes the MI between $q(h,d)$ and $q(h)q(d)$, and enforces the extraction of DLR features which are not redundant with HCR features. The importance of the MI in the loss function is weighted by κ, which we empirically set to 1 (see Sect. A.2 of the appendix for more details). To ensure that

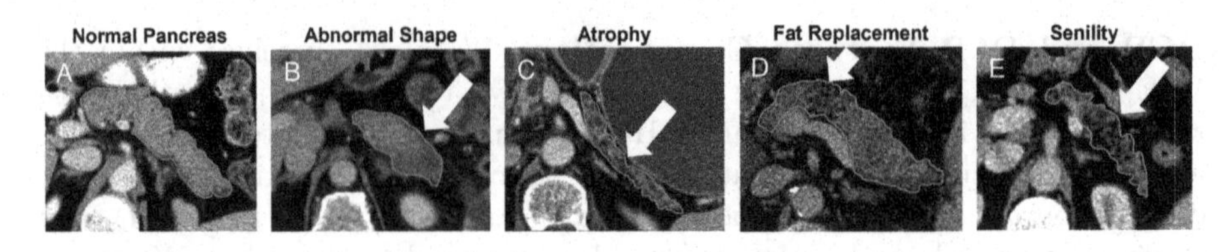

Fig. 2. Portal CT scans showing early markers of pancreatic cancer. Pancreas are delineated in orange. (A) shows a normal pancreas. White arrows indicate an abnormal enlarged tail (B), a parenchymal atrophy (C), fat replacement in the neck of the pancreas (D) and senile characteristics (E). (Color figure online)

the density-ratio is well-estimated, as explained in [8], we opt for an alternate optimization scheme between the VAE model and the discriminator $\mathcal{D}_\lambda$: every 5 epochs, we freeze the optimization of the VAE, train the discriminator for 150 epochs, and continue the optimization of the VAE model.

Early Cancer Markers Prediction. Once the VAE model is trained, DLR can be extracted and leveraged to predict cancer markers. We propose to train, for each marker of interest, a classifier $\mathcal{C}_M$ based on the concatenation of HCR and DLR extracted by our model. Unlike VAE training, which is unsupervised and task-agnostic, $\mathcal{C}_M$ training is supervised and specific to a cancer marker.

3 Experiments

We illustrate our method on the pancreas, for which we aim to predict four early markers of abnormality that manifest prior to the onset of visible lesions:

(i) *Abnormal shape:* Changes in the shape of the pancreas can be associated with pancreatic cancer as the tumor growth can lead to various structural changes in the pancreas [15,25];
(ii) *Atrophy:* Pancreatic atrophy may signal pancreatic cancer [19] and can indicate small isodense lesions [26];
(iii) *Fat replacement:* Fat replacement is characterized by the accumulation of fat within the pancreas and is associated with various metabolic diseases, pancreatitis, pancreatic cancer, and precancer [7,17,19]. While this mainly modifies the texture, severe fat replacement can also affect the shape by inducing lobulated margins;
(iv) *Senility:* Anatomical changes in the pancreas, such as pancreatic atrophy, fatty replacement and fibrosis have been documented in elderly individuals and increase the susceptibility of individuals to pancreatic cancer [7,18].

These early signs are illustrated in Fig. 2.

Dataset. Data were obtained from our private cohort and split into two independent datasets $\mathcal{D}^{Train}$ and $\mathcal{D}^{Test}$, containing 2319 and 1094 abdominal portal CT scans from six and three independent medical centers, respectively. The reference labels regarding the early markers previously described were obtained based on the assessment of the CT scan by a pool of 7 radiologists. Reference labels were collected for 676 cases of $\mathcal{D}^{Train}$ and all the subjects from $\mathcal{D}^{Test}$.

Preprocessing. For all the subjects, pancreas segmentation masks were obtained using a segmentation model derived from the nnU-Net [6] and manually reviewed by radiologists. The CT images and corresponding masks were resampled to $1 \times 1 \times 2$ mm^3 in the (x, y, z) directions, and centered in a volume of size $192 \times 128 \times 64$ voxels. Images intensities were clipped to the $[0.5, 99.5]$ percentiles and standardized based on the percentiles, mean and standard deviation of the pancreas intensities in $\mathcal{D}^{Train}$.

Extracting HCR and DLR. 32 HCR features were extracted utilizing the pyradiomics library [24], focusing exclusively on shape and first-order intensity features (see Sect. A.3 in the appendix for the comprehensive list). Complementary DLR features were extracted using the VAE model of Sect. 2. The architecture followed the U-Net [22] encoder-decoder scheme without skip connections. The number of convolutional layers and the convolutional blocks were automatically inferred thanks to the nnU-Net self-configuring procedure [6] (see Sect. A.4 in the appendix for details). The model was trained on $\mathcal{D}^{Train}$ for 1000 epochs. The dimension of DLR features d was set to 32, resulting in a final latent space dimension for the VAE of 64. Data augmentation consisting of rotation and cropping was applied during training.

Predicting Early Cancer Markers. For each marker, a logistic regression was trained based on the concatenation of HCR and DLR features extracted from the subjects in $\mathcal{D}^{Train}$ for whom reference labels were available. The logistic regression was regularized using $\mathcal{L}_2$ penalty, with a default regularization coefficient of 1. Final predictions for $\mathcal{D}^{Test}$ were derived by ensembling models obtained through a four-fold cross-validation setup.

4 Results

Quantitative Results. To demonstrate the usefulness of extracting DLR with MI minimization, two VAEs were trained. Both followed the same procedure (detailed in Fig. 1) but differed only in the presence or absence of the MI minimization term in their loss function. Then, several logistic regression models with different inputs were trained in order to assess the effect of combining HCR and DLR features. In total, the following experiments were run:

- **HCR only:** H_{32} **and** H_{64}. These two experiments use the 32 basic HCR features described in Sect. A.3 of the appendix, and H_{64} uses a further 32 HCR gray-level features calculated by the pyradiomics library [24] and selected by recursive feature elimination.
- **DLR only:** $\mathrm{D}_{32}^{\mathrm{MI}}$ **and** D_{32}. 32 DLR features extracted by a VAE with and without MI minimization, respectively;
- **HCR + DLR:** $\mathrm{HD}_{64}^{\mathrm{MI}}$ **and** HD_{64}. 32 basic HCR features + 32 DLR features extracted by a VAE with and without MI minimization, respectively.

Thus, the logistic regressions of H_{32}, D_{32} and $\mathrm{D}_{32}^{\mathrm{MI}}$ used vectors of size 32, while those of H_{64}, HD_{64} and $\mathrm{HD}_{64}^{\mathrm{MI}}$ used vectors of size 64. Prediction results for each of the four cancer markers are presented in Table 1.

Table 1. Pancreatic cancer marker prediction. For each experiment, we report the means and standard deviations of the AUC (in %) obtained by bootstrapping with 10000 repetitions. For each line, first and second best results are in bold and underlined, respectively. The last row shows the difference in AUC compared with H_{32}, averaged over the different markers. DLR and HCR refer to Deep Learning Radiomics and Hand-Crafted Radiomics, respectively.

	HCR only		DLR only		HCR + DLR	
	H_{32}	H_{64}	D_{32}	$\mathrm{D}_{32}^{\mathrm{MI}}$	HD_{64}	$\mathrm{HD}_{64}^{\mathrm{MI}}$
Abnormal Shape	68.38± 0.07	68.11± 0.07	67.66± 0.07	**72.41± 0.07**	<u>71.2± 0.07</u>	70.07± 0.07
Atrophy	81.05± 0.06	<u>81.57± 0.05</u>	74.08± 0.07	79.08± 0.06	80.82± 0.06	**82.57± 0.06**
Fat Replacement	<u>70.55± 0.07</u>	69.78± 0.08	65.96± 0.08	65.74± 0.07	69.28± 0.08	**71.05± 0.07**
Senility	71.63± 0.08	70.21± 0.08	70.18± 0.07	69.1± 0.08	<u>72.28± 0.08</u>	**72.44± 0.07**
δ w.r.t H_{32}	–	-0.48± 0.07	-3.43± 0.07	-1.32± 0.07	<u>0.49</u>± 0.07	**1.13± 0.07**

The comparison between H_{32} and H_{64} showed that adding 32 gray-level HCR features was not beneficial as results were similar, or even decreased: for instance, for senility, the AUC went from 71.63 % (H_{32}) to 70.21 % (H_{64}). On average, the AUC of H_{64} lost -0.48 points compared with H_{32}. These experiments demonstrated the power of the 32 basic HCR features, and the need to find complementary features that would add value.

Then, for almost all markers, H_{32} outperformed D_{32} and $\mathrm{D}_{32}^{\mathrm{MI}}$, meaning that no VAE, whether trained with or without MI minimization, managed to automatically extract 32 DLR features as informative as the 32 basic HCR features used by H_{32}. For texture-related markers, such as fat replacement and senility, MI minimization did not produce clear differences. On the other hand, on shape-related markers, the DLR features learned by $\mathrm{D}_{32}^{\mathrm{MI}}$ were shown to be more relevant than those learned by D_{32} with a basic VAE. Thus, on average, DLR features were better when extracted by a VAE trained with MI minimization, but still proved less informative than HCR features.

Finally, experiments HD_{64} and $\mathrm{HD}_{64}^{\mathrm{MI}}$ showed that combining the two types of radiomics is beneficial since the average AUC gained 0.49 (HD_{64}) and 1.13 %

(HD_{64}^{MI}) compared to H_{32}. Yet, results demonstrated that minimizing the redundancy produced the best results compared with all other approaches. Indeed, in HD_{64}, adding 32 DLR features produced variable results depending on the markers: compared to H_{32}, the AUC increased by a maximum of 2.82% for abnormal shape prediction, and dropped by a maximum of 1.27% for predicting fat replacement. On the other hand, HD_{64}^{MI} outperformed H_{32} on all prediction problems, meaning that the non-redundant DLR features systematically provided useful information.

Influence of the Latent Space. To explore the influence of the latent space dimension on the prediction performances, we replicated the HD_{64}^{MI} experiment with increasing size L of the latent space, and reported prediction results in Table 2. Table 2 shows that increasing the latent space size resulted in lower classification performances. Specifically, a latent space size of 32 provided the most relevant DLR features.

Table 2. Pancreatic cancer marker prediction with varying latent space size. For each experiment, a VAE with Mutual Information (MI) minimization and latent space size L was trained. Predictions were obtained after training logistic regressions on 32 basic HCR features + L DLR features extracted by a VAE with MI minimization. We report the means and standard deviations of the AUC (in %) obtained on the test set by bootstrapping with 10000 repetitions. For each line, first best results are in bold. DLR and HCR refer to Deep Learning Radiomics and Hand-Crafted Radiomics, respectively.

	$L = 32$	$L = 64$	$L = 256$	$L = 512$	$L = 1024$	$L = 2048$
Abnormal Shape	**70.07± 0.07**	69.02± 0.07	68.87± 0.07	69.91± 0.07	69.33± 0.07	68.68± 0.07
Atrophy	82.57± 0.06	82.28± 0.05	81.77± 0.06	**82.68± 0.05**	80.9± 0.06	80.21± 0.06
Fat Replacement	**71.05± 0.07**	70.91± 0.07	70.23± 0.08	70.45± 0.08	69.55± 0.07	68.96± 0.08
Senility	**72.44± 0.07**	72.02± 0.07	70.38± 0.08	71.65± 0.08	72.03± 0.07	69.6± 0.08

Qualitative Results. To visualize the effect of the extracted DLR features, we looked at the absolute value of the logistic regression weights for D_{32} and D_{32}^{MI} in two ways. In Fig. 3-A, the absolute value of these coefficients are displayed. The higher the absolute value of the coefficient, the higher its importance in the logistic regression prediction. When the MI was not minimized, HCR features had stronger importance than DLR ones. On the other hand, when we encouraged the independence between the two types of features through MI minimization, the contribution of DLR features to the prediction increased. Figure 3-B shows the number of DLR features among the k features with highest importance, for increasing values of k. HD_{64}^{MI} and HD_{64} are shown in blue and orange, respectively. In addition, two extreme scenarios are shown: one where the logistic regression is predominantly influenced by the DLR features (in green), and another one where the logistic regression is primarily driven by the HCR features (in red). We can see that the blue curve approached the green curve, meaning that DLR features from HD_{64}^{MI} contributed more to the outcome prediction. When the MI was not minimized, DLR features had less influence on the predictions as the orange curve approached the scenario in which DLR would be ignored.

Reconstruction Performances. To explore the reconstruction performances of the VAE, we computed the average l_2 error per voxel between the original test images and their corresponding reconstructions. Upon applying nnU-Net's [6] automatic intensity normalization procedure, voxel intensities were observed to range from -3 to 2.3. Specifically, we employed a VAE with a latent space dimension of $L = 32$ and MI minimization during training. The resulting reconstruction error was found to be $(4.4 \pm 1.4) \times 10^{-3}$, which was comparable to the l_2 error obtained from a VAE trained without MI minimization, amounting to $(4.1 \pm 1.4) \times 10^{-3}$. These observations suggest that the introduction of MI minimization did not significantly impact the quality of the reconstructed images, neither resulting in deterioration nor improvement. Additionally, Table 3

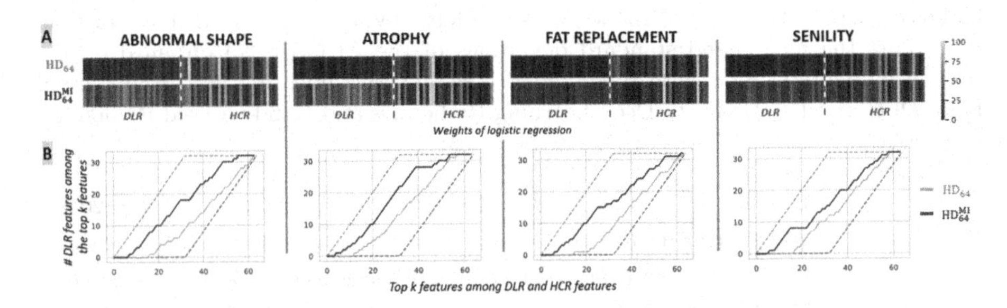

Fig. 3. Qualitative assessment of the Deep Learning Radiomics (DLR) and Hand-Crafted Radiomics (HCR) features through the coefficients of the logistic regressions. *A: Absolute value of the coefficients of the logistic regressions.* We plot, for each logistic regression corresponding to one marker, the absolute value of the coefficient for each of the 64 features. The first 32 features corresponded to DLR, while the 32 remaining features corresponded to HCR. *B: Number of DLR features among the top k features.* Dashed lines represent the extreme scenarios in which all 32 DLR are more informative than all 32 HCR (green) or all 32 HCR are more informative than all 32 DLR (red). (Color figure online)

Table 3. Reconstruction performances with varying latent space sizes. For each experiment, a VAE with Mutual Information minimization and latent space size L was trained. We report the l_2 error per voxel between the original image and its reconstruction, with voxel intensities varying in $[-3, 2.3]$.

	$L = 32$	$L = 64$	$L = 256$	$L = 512$	$L = 1024$	$L = 2048$
l_2 error $\times 10^3$	4.4± 1.4	4.4± 1.4	4.4± 1.4	4.4± 1.4	4.3± 1.4	4.3± 1.5

further explores the relationship between reconstruction performance and latent space sizes, demonstrating that increasing the latent space size did not have a discernible effect on the quality of the reconstructions.

5 Discussion and Conclusion

We presented a method to learn DLR features that are not redundant with HCR ones. The method was based on the well-known VAE framework [9] that extracted DLR features from masked images in an unsupervised manner. The complementarity between the two types of radiomics features was enforced by minimizing their MI, and the resulting features were used to train classifiers predicting different cancer markers. Experiments in the case of four early markers of pancreatic cancer indicated that our method increased prediction performances with respect to two state-of-the-art approaches. These findings suggest that our approach holds potential to improve patient survival outcomes. Qualitative results confirmed the advantages of minimizing the MI during training, as it resulted in the generation of DLR features that were complementary to HCR features and more prominently utilized for marker prediction. These results were obtained on a large and independent test set, which is particularly important as radiomics models require robust validation strategies to ensure their generalization and reproducibility when applied to new datasets [1]. With this in mind, it might be interesting to further encourage this feature efficiency by imposing independence between the DLR features themselves. Another research avenue could be to simplify the proposed pipeline by developing an end-to-end network capable of performing both feature extraction and classification tasks within a unified framework. Achieving this objective would necessitate the simultaneous training of the feature extractor and multiple sub-networks for each classification task. However, this approach might pose challenges in terms of training complexity, particularly due to the presence of substantial class imbalances across the various classification tasks. Alternatively, another possibility is to train an end-to-end convolutional neural network (CNN). Although more direct in nature, this approach would entail the training of a separate CNN for each question, which could be computationally heavier compared to the calibration of a logistic regression based on a single feature extractor, as suggested in our current work. Future studies should also address the interpretability of the extracted DLR features, as this aspect was not covered in the present work.

Acknowledgments. This work was partly funded by a CIFRE grant from ANRT # 2020/1448.

A Appendix

A.1 Estimating the Mutual Information

The Mutual Information (MI) is estimated following the density-ratio trick [8] which requires to train a discriminator $\mathcal{D}_\lambda$ predicting whether concatenated radiomics vectors $[h, d]$ come from $q(h, d)$ or $q(h)q(d)$. Samples for training $\mathcal{D}_\lambda$ are obtained following the procedure shown in Fig. 4. In practice, $\mathcal{D}_\lambda$ is modeled as a 2-layer Multi Layer Perceptron with ReLu activation, which is trained by minimizing a binary cross-entropy (BCE) loss term. Once the discriminator is trained, the MI between HCR and DLR features can be approximated as follows:

$$\mathrm{MI}(h, d) = \mathbb{E}_{q(h,d)}\left[\log \frac{q(h, d)}{q(h)q(d)}\right] \approx \sum_i \mathrm{ReLU}\left(\left[\log \frac{\mathcal{D}_\lambda(h_i, d_i)}{1 - \mathcal{D}_\lambda(h_i, d_i)}\right]\right). \quad (5)$$

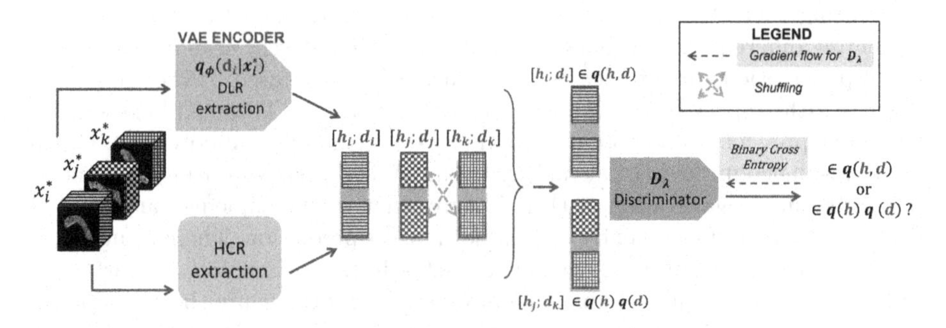

Fig. 4. Training $\mathcal{D}_\lambda$. Given three different input images x_i^*, x_j^* and x_k^*, the corresponding HCR and DLR features are computed: h_j, h_j, h_k and d_i, d_j, d_k. Samples from $q(h, d)$ are obtained by concatenating features of a same image (h_i and d_i for instance), while samples from $q(h)q(d)$ are obtained by concatenating h_k and d_j with $k \neq j$.

A.2 Influence of the Hyperparameter κ

The final loss function for training our model is:

$$\mathcal{L} = \mathcal{L}_{\mathrm{VAE}} + \kappa KL[q(h, d) \mid q(h)q(d)] \quad (6)$$

where κ is a hyperparameter weighting the importance of the the mutual information in the total loss function. Table 4 reports prediction results obtained with different values of κ. According to these results, κ was set to 1 in all our experiments.

Table 4. Cancer marker prediction scores for different values of κ. For each experiment, we report the means and standard deviations of the AUC (in %) obtained by bootstrapping with 10000 repetitions. For each line, best result is in bold.

	$\kappa = 0.01$	$\kappa = 0.1$	$\kappa = 1$	$\kappa = 10$
General Shape	70.44± 0.07	70.01± 0.07	70.07± 0.07	**71.03± 0.07**
Atrophy	80.82± 0.05	81.43± 0.06	**82.57± 0.06**	80.77± 0.06
Fat Replacement	69.52± 0.08	70.5± 0.07	**71.05± 0.07**	68.65± 0.08
Senility	**73.14± 0.08**	72.36± 0.08	72.44± 0.07	72.38± 0.08

A.3 HCR Features Extraction

32 HCR features were extracted using the pyradiomics library [24]:

- **14 shape features** describing the size and shape of the pancreas
  - Mesh Volume
  - Voxel Volume
  - Surface Area
  - Surface Area to Volume ratio
  - Sphericity
  - Maximum 3D diameter
  - Maximum 2D diameter in the axial plane
  - Maximum 2D diameter in the coronal plane
  - Maximum 2D diameter in the sagittal plane
  - Major Axis Length
  - Minor Axis Length
  - Least Axis Length
  - Elongation
  - Flatness
- **18 first-order intensity features** describing the intensities distribution within the organ
  - Energy
  - Total Energy
  - Entropy
  - Minimum
  - 10th percentile
  - 90th percentile
  - Maximum
  - Mean
  - Median
  - Interquartile Range
  - Range
  - Mean Absolute Deviation
  - Robust Mean Absolute Deviation
  - Root Mean Squared
  - Skewness
  - Kurtosis
  - Variance
  - Uniformity

More details about each feature can be found on the online documentation.

A.4 Model Architecture

As detailed in Fig. 5, the proposed variational autoencoder (VAE) followed a 3D encoder-decoder architecture. The network topology (number of convolutions per block, filter sizes) was chosen based on the nnU-Net self-configuring procedure [6], resulting in 1, 110, 240 trainable parameters. The VAE was trained on 1000 epochs with a batch size of size 32. Every 5 epochs, the VAE was frozen and the discriminator $\mathcal{D}_\lambda$ was trained for 150 epochs with a batch size equal to the total training dataset. The VAE and $\mathcal{D}_\lambda$ were optimized using two independent Adam optimizers with a learning rate of 10^{-3}.

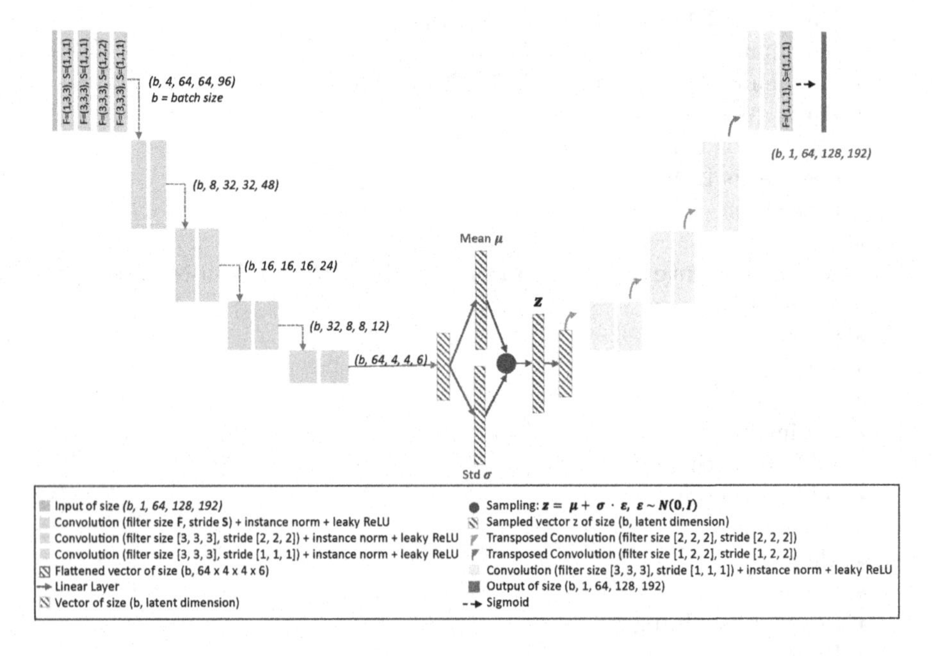

Fig. 5. Architecture of the proposed VAE

References

1. Aerts, H.J., et al.: Decoding tumour phenotype by noninvasive imaging using a quantitative radiomics approach. Nat. Commun. **5**(1), 4006 (2014)
2. Afshar, P., Mohammadi, A., Plataniotis, K.N., Oikonomou, A., Benali, H.: From handcrafted to deep-learning-based cancer radiomics: challenges and opportunities. IEEE Signal Process. Maga. **36**(4), 132–160 (2019)
3. Antropova, N., Huynh, B.Q., Giger, M.L.: A deep feature fusion methodology for breast cancer diagnosis demonstrated on three imaging modality datasets. Med. Phys. **44**(10), 5162–5171 (2017)
4. Chen, S., et al.: Automatic scoring of multiple semantic attributes with multi-task feature leverage: a study on pulmonary nodules in CT images. IEEE Trans. Med. Imaging **36**(3), 802–814 (2016)

5. Huynh, B.Q., Li, H., Giger, M.L.: Digital mammographic tumor classification using transfer learning from deep convolutional neural networks. J. Med. Imaging **3**(3), 034501–034501 (2016)
6. Isensee, F., et al.: nnU-Net: a self-configuring method for deep learning-based biomedical image segmentation. Nat. Methods **18**(2), 203–211 (2021)
7. Khoury, T., Asombang, A.W., Berzin, T.M., Cohen, J., Pleskow, D.K., Mizrahi, M.: The clinical implications of fatty pancreas: a concise review. Digest. Dis. Sci. **62**, 2658–2667 (2017)
8. Kim, H., Mnih, A.: Disentangling by factorising. In: International Conference on Machine Learning, pp. 2649–2658. PMLR (2018)
9. Kingma, D.P., Welling, M.: Auto-encoding variational bayes. In: 2nd International Conference on Learning Representations, ICLR (2014)
10. Kumar, D., Wong, A., Clausi, D.A.: Lung nodule classification using deep features in CT images. In: Conference on Computer and Robot Vision, pp. 133–138. IEEE (2015)
11. Kumar, V., et al.: Radiomics: the process and the challenges. Magn. Reson. Imaging **30**(9), 1234–1248 (2012)
12. Lambin, P., et al.: Radiomics: extracting more information from medical images using advanced feature analysis. Eur. J. Cancer **48**(4), 441–446 (2012)
13. Lao, J., et al.: A deep learning-based radiomics model for prediction of survival in glioblastoma multiforme. Sci. Rep. **7**(1), 10353 (2017)
14. Litjens, G., et al.: A survey on deep learning in medical image analysis. Med. Image Anal. **42**, 60–88 (2017)
15. Liu, F., Xie, L., Xia, Y., Fishman, E., Yuille, A.: Joint shape representation and classification for detecting PDAC. In: Suk, H.-I., Liu, M., Yan, P., Lian, C. (eds.) MLMI 2019. LNCS, vol. 11861, pp. 212–220. Springer, Cham (2019). https://doi.org/10.1007/978-3-030-32692-0_25
16. Liu, S., Xie, Y., Jirapatnakul, A., Reeves, A.P.: Pulmonary nodule classification in lung cancer screening with three-dimensional convolutional neural networks. J. Med. Imaging **4**(4), 041308–041308 (2017)
17. Majumder, S., Philip, N.A., Takahashi, N., Levy, M.J., Singh, V.P., Chari, S.T.: Fatty pancreas: should we be concerned? Pancreas **46**(10), 1251 (2017)
18. Matsuda, Y.: Age-related morphological changes in the pancreas and their association with pancreatic carcinogenesis. Pathol. Int. **69**(8), 450–462 (2019)
19. Miura, S., et al.: Focal parenchymal atrophy and fat replacement are clues for early diagnosis of pancreatic cancer with abnormalities of the main pancreatic duct. Tohoku J. Exp. Med. **252**(1), 63–71 (2020)
20. Paul, R., et al.: Deep feature transfer learning in combination with traditional features predicts survival among patients with lung adenocarcinoma. Tomography **2**(4), 388–395 (2016)
21. Ravì, D., et al.: Deep learning for health informatics. J. Biomed. Health Inf. **21**(1), 4–21 (2016)
22. Ronneberger, O., Fischer, P., Brox, T.: U-net: convolutional networks for biomedical image segmentation. In: Navab, N., Hornegger, J., Wells, W.M., Frangi, A.F. (eds.) MICCAI 2015. LNCS, vol. 9351, pp. 234–241. Springer, Cham (2015). https://doi.org/10.1007/978-3-319-24574-4_28
23. Shafiee, M.J., Chung, A.G., Khalvati, F., Haider, M.A., Wong, A.: Discovery radiomics via evolutionary deep radiomic sequencer discovery for pathologically proven lung cancer detection. J. Med. Imaging **4**(4), 041305–041305 (2017)
24. Van Griethuysen, J.J., et al.: Computational radiomics system to decode the radiographic phenotype. Cancer Res. **77**(21), e104–e107 (2017)

25. Vétil, R., et al.: Learning shape distributions from large databases of healthy organs: applications to zero-shot and few-shot abnormal pancreas detection. In: International Conference on Medical Image Computing and Computer Assisted Intervention, Part II, pp. 464–473. Springer, Heidelberg (2022). https://doi.org/10.1007/978-3-031-16434-7_45
26. Yamao, K., et al.: Partial pancreatic parenchymal atrophy is a new specific finding to diagnose small pancreatic cancer ($\leq$ 10 mm) including carcinoma in situ: comparison with localized benign main pancreatic duct stenosis patients. Diagnostics **10**(7), 445 (2020)

Assessing the Performance of Deep Learning-Based Models for Prostate Cancer Segmentation Using Uncertainty Scores

Pablo Cesar Quihui-Rubio[1(✉)], Daniel Flores-Araiza[1], Gilberto Ochoa-Ruiz[1], Miguel Gonzalez-Mendoza[1], and Christian Mata[2,3]

[1] School of Engineering and Sciences, Tecnologico de Monterrey, Monterrey, Mexico
pabloqulhul@gmail.com, gilberto.ochoa@tec.mx
[2] Universitat Politècnica de Catalunya, 08019 Barcelona, Catalonia, Spain
[3] Pediatric Computational Imaging Research Group, Hospital Sant Joan de Déu, Esplugues de Llobregat, 08950 Catalonia, Spain

Abstract. This study focuses on comparing deep learning methods for the segmentation and quantification of uncertainty in prostate segmentation from MRI images. The aim is to improve the workflow of prostate cancer detection and diagnosis. Seven different U-Net-based architectures, augmented with Monte-Carlo dropout, are evaluated for automatic segmentation of the central zone, peripheral zone, transition zone, and tumor, with uncertainty estimation. The top-performing model in this study is the Attention R2U-Net, achieving a mean Intersection over Union (IoU) of 76.3% $\pm$ 0.003 and Dice Similarity Coefficient (DSC) of 85% $\pm$ 0.003 for segmenting all zones. Additionally, Attention R2U-Net exhibits the lowest uncertainty values, particularly in the boundaries of the transition zone and tumor, when compared to the other models.

Keywords: Segmentation · Uncertainty Quantification · Prostate · Cancer · Deep Learning · Computer Vision

1 Introduction

Prostate cancer (PCa) is the most common solid non-cutaneous cancer in men and is among the most common causes of cancer-related deaths in 13 regions of the world [1]. According to a recent overview, in 2020 prostate cancer was the most frequently diagnosed cancer in males in 12 regions of the world, which translates to around 1.41 million new cases [1]. However, when detected in early stages, the survival rate for regional PCa is almost 100%. In contrast, the survival rate when the cancer is spread to other parts of the body is of only 30% [2].

Magnetic Resonance Imaging (MRI) is the most widely available non-invasive and sensitive tool for detection, localization and staging of PCa, due to its high resolution, excellent spontaneous contrast of soft tissues, and the possibility of

© The Author(s), under exclusive license to Springer Nature Switzerland AG 2023
S. Ali et al. (Eds.): CaPTion 2023, LNCS 14295, pp. 83–93, 2023.
https://doi.org/10.1007/978-3-031-45350-2_7

multi-planar and multi-parametric scanning [3]. MRI can be also be used for PCa detection through the segmentation of Regions of Interest (ROI). The use of image segmentation for PCa can help determine the localization and the volume of the cancerous tissue [4]. Although prostate image segmentation is a relatively old problem and some novel methods have been proposed, radiologists still perform a manual segmentation of the prostate gland and regions of interest (central zone, peripheral zone, and transition zone) [5]. This manual process is time-consuming, and is sensitive to the specialist experience, resulting in a significant intra- and inter-specialist variability. Therefore, automating the process of segmentation of prostate and gland regions of interest, may help save time for practitioner radiologists and additionally can be used as a training tool for others. One of the most popular architectures is the U-Net [6] model, which has been the inspiration behind many recent works in the literature, such as Swin U-Net [7], or R2U-Net [8]. While these models have yielded positive outcomes, inconsistencies in performance have been observed in U-Net-based segmentation due to the prostate's anatomical structure. The boundaries between zones can distort semantic features, leading to unreliable results. Furthermore, automatic segmentation typically produces deterministic segmentation outcomes [9], and there is insufficient information available about the model's confidence level [10]. Despite their successes in many medical image analysis applications, DL algorithms are usually not translated into real-world clinical scenarios because these do not provide information about the uncertainty associated with their prediction. This is problematic in the challenging context of pathological structures segmentation (e.g., tumors) as even the top-performing methods are prone to errors, and due to the lack of uncertainty information, it results impossible tell apart different sorts of erroneous predictions.

Therefore, the overall segmentation workflow can be improved by providing the uncertainties of the model that could allow end-users (e.g., clinicians) to review and refine cases with high uncertainty.

In this work, we carry out a thorough assessment of automatic prostate zone segmentation models using U-Net, Attention U-Net, Dense U-Net, Attention Dense U-Net, R2U-Net, Attention R2U-Net, and Swin U-Net architectures. Additional to the segmentation task, we include the pixel-wise estimation of the uncertainty, which can be done by obtaining a probability distribution of the weights of the model. The zones evaluated in this work are the central zone (CZ), the peripheral zone (PZ), transition zone (TZ), and, in the case of a disease, the tumor zone (TUM), unlike previous works which only evaluate CZ and PZ [10].

This paper has five sections including this introduction. Section 2 provides a review about what has been done in previous works related to prostate segmentation and uncertainty quantification. Section 3 the dataset used is described, followed by a description of the uncertainty quantification procedure in this segmentation task. In Sect. 4 the results of the experiments are discussed in detail. Finally the conclusion of this work is presented in Sect. 6.

2 Related Work

2.1 Deep Learning Segmentation

For segmentation, one of the best known models in the literature is the U-Net architecture [6], which is the base for many other novel models. The work from Zhu et *al.* [11] proposes a U-Net based network to segment the whole prostate gland, obtaining encouraging results (DSC of 0.885). Moreover, this architecture has served as the inspiration for some variants that enhance the performance of the original model. One example is the work from Clark et *al.* [12] that presents a model that combines concepts from the U-Net and the inception architectures. Another example is the work presented by Oktay et *al.* [13], which proposes the addition of attention gates inside the original U-Net model with the intention of focusing on specific target structures. The addition of attention has served as base for other architectures such as Attention Dense U-Net [14], Attention R2U-Net [8], among others. Also, the introduction of Transformers in U-Net architectures is a novel approach for segmentation task that had demonstrated a good performance in biomedical images, such as Swin U-Net [7]. Despite this, during the course of this study, no other research was found that segmented the four zones discussed in this paper. Therefore, the number of studies that consider a third zone (TZ) is still limited, this is more likely because the most common datasets used are PROMISE-12 and the one from the PROSTATEx challenge, with only CZ and PZ. In addition to that, providing a value that quantifies the uncertainty of the predictions can improve the overall workflow since it could easily allow refining uncertain cases by human experts.

2.2 Uncertainty Quantification

The work from Theckel et *al.* [15] introduces a U-Net architecture with spatial dropout to measure the uncertainty related to the segmentation of macular degeneration, utilizing different sizes of input data. The work from Suman et *al.* [16] applied the uncertainty quantification problem to retinal imaging using a ResNet-based model, modified with standard random dropout layers before every convolutional block. The work from Liu et *al.* [10] proposes an automatic segmentation of the prostate zones and introduces a pixel-wise uncertainty estimator using a ResNet50 backbone with attention and dropout layers.

3 Materials and Methods

3.1 Dataset

The dataset used in the present work was provided by *Universidad Politécnica de Cataluña* (UPC) in Barcelona, and Centre Hospitalaire de Dijon in France. The dataset consists of three-dimensional T2-weighted fast spin-echo (TR/TE/ETL: 3600 ms/ 143 ms/109, slice thickness: 1.25 mm) images acquired with sub-millimeter pixel resolution in an oblique axial plane. The number of patients in the dataset are 19, with a total of 205 images with their corresponding annotation masks (of prostate zones) used as ground truth which were validated by experts using a dedicated tool [17].

The full dataset of 205 images, contains four different combination of zones, being: (CZ+PZ), (CZ+PZ+TZ), (CZ+PZ+Tumor), and (CZ+PZ+TZ+Tumor) with 73, 68, 23, and 41 images, respectively. For the purpose of this work, the dataset was divided in 85% for training and 15% for testing.

3.2 Uncertainty Estimation in Prostate Segmentation

Epistemic and aleatory uncertainties are the two major types of uncertainty that can be quantified. Epistemic uncertainty captures the uncertainty related to the models parameters caused by the lack of data, and, aleatory uncertainty captures the noise inherent in the input data [10]. The sum of both uncertainties forms the predictive uncertainty.

In this work, the uncertainty of seven different U-net-based models was measured in the test set. To approximate the inference of a model, Monte Carlo (MC) dropout of a hidden layer was performed. MC Dropout is a technique used in neural networks to incorporate uncertainty. It treats a network with dropped-out neurons as Monte Carlo samples from all possible combinations, approximating a Gaussian process [10,18]. The minimization of cross-entropy loss is similar to minimizing the divergence of the predicted distribution [16]. Using MC Dropout, pixel-wise epistemic uncertainty can be computed as a variational Bayesian inference problem [16]. During predictions or testing, dropout is also necessary. The main focus of this study is to investigate the predictive uncertainty of prostate segmentation, which can be quantified using the entropy of the predictive distribution [10].

3.3 Proposed Work

This work uses the original U-Net model and six U-Net extensions from the literature: Attention U-Net [13], Dense U-Net [19], Attention Dense U-Net [14], R2U-Net [8], Attention R2U-Net, and Swin U-Net [7]. These architectures had demonstrated great performance segmenting biomedical images, even some of them with public prostate's datasets including CZ and PZ. However, unlike in other works, we proposed to compare the performance segmenting the three main zones of the prostate (CZ, PZ, and TZ) and a tumor tissue if it is present, using the dataset described in Sect. 3.1.

Before the final training, an hyperparameter tunning proccess using a stratified 5-Fold validation with the training set was carried out using the base U-Net model in order to obtain the optimal combination of data augmentation, learning rate and an approximation of epochs for training. The results demonstrated that including data augmentation in the training did not increase significantly the performance of the models. Therefore we decided to use the original dataset without data augmentation due to computational resources and time processing. The previously mentioned models were trained for 145 epochs, using Adam optimizer with a learning rate of 1e–4 and Categorical Cross-Entropy (CCE) loss function. The performance was evaluated using Dice Score (DSC) and Intersection over Union (IoU) as the main metrics.

4 Results and Discussion

4.1 Quantitative Results

Table 1 shows a summary of evaluation results of the seven studied architectures, in terms of two metrics (DSC and IoU) and loss value. In order to obtain these results, the evaluation of each model was performed $T = 50$ times, and due to the incorporation of MC Dropouts the results were different each time. Therefore, the average of all evaluations and prostate zones is reported with their corresponding standard deviation.

Table 1. Comparison of model performance in segmentation metrics and loss value. The metrics are denoted by upward (↑) or downward (↓) arrows, indicating the desired direction of values. Bold values highlighted in green represent the best score achieved among all models.

Model	IoU ↑	DSC ↑	Loss ↓
U-Net	0.676 ± 0.021	0.770 ± 0.021	0.0139 ± 0.0007
Attention U-Net	0.688 ± 0.011	0.781 ± 0.010	0.0132 ± 0.0003
Swin U-Net	0.725 ± 0.014	0.816 ± 0.014	0.0134 ± 0.0002
Dense U-Net	0.754 ± 0.004	0.846 ± 0.004	0.0146 ± 0.0003
Attention Dense U-Net	0.760 ± 0.006	0.847 ± 0.005	0.0154 ± 0.0004
R2U-Net	**0.764 ± 0.002**	**0.850 ± 0.002**	0.0119 ± 0.0001
Attention R2U-Net	0.763 ± 0.003	**0.850 ± 0.003**	**0.0113 ± 0.0001**

Based on the metrics values, it can be seen that U-Net was the model with worst performance. The use of attention to focus on the ROI helped to slightly outperform the performance in segmentation tasks compared to the original U-Net by around 1–2% for IoU and DSC.

Moving to Swin U-Net, a novel architecture from the state-of-the-art that uses Swin Transformers [7] achieved to increase the IoU and DSC values by more than 7%, and lower loss value compared to U-Net.

In the case of Dense U-Net, the performance of the model exceeds the previous three architectures, with IoU and DSC scores 11% and 10% better than the base U-Net, respectively, with a loss value of 0.0146. As a plus, this model did not need more computational resources or time during its training compared to base U-Net. The next model consisted on the incorporation of attention modules to Dense U-Net, which again outperformed all the previous models in the segmentation metrics by 12% of IoU, and 10% of DSC compared to U-Net. However, it achieved the higher loss value among all of 0.0154.

The last two architectures R2U-Net and Attention R2U-Net achieved very similar results, but outperformed all the other models with values of 76.4% and 85% for IoU and DSC, respectively, and the lowest loss value of 0.0113 for the Attention R2U-Net.

As mentioned before, an uncertainty comparison between the architectures was carried out per each prostate zone, as well as for the full image with its corresponding standard deviation as it is shown in Fig. 1. The results shown in this figure can help us to determine, in relation with previous table, which model achieved to segment with more certain the prostate and its zones.

In Fig. 1 it is observed that overall, the model that had the lowest mean uncertainty segmenting all the images in the test set was R2U-Net with a mean value of 0.048 ± 0.014 after 50 predictions, validating the results obtained in the Table 1, being the most reliable and accurate model overall thanks to the use of recurrent and residual units to get more context information.

Furthermore, the Attention U-Net was the one with the highest uncertainty overall with a value of 0.086 ± 0.023, having poor results in comparison to the other models. U-Net and Swin U-Net obtained very similar results in most of the prostate zones, although in the case of the TZ and Tumor, Swin U-Net achieved lower uncertainty.

Dense U-Net, Attention Dense U-Net and Attention R2U-Net succeeded in obtaining smaller uncertainty mean values than U-Net (0.055 ± 0.018, 0.054 ± 0.018, and 0.052 ± 0.014, respectively). Although, TZ and Tumor are the zones less present in the dataset, and where it looks to be more complex

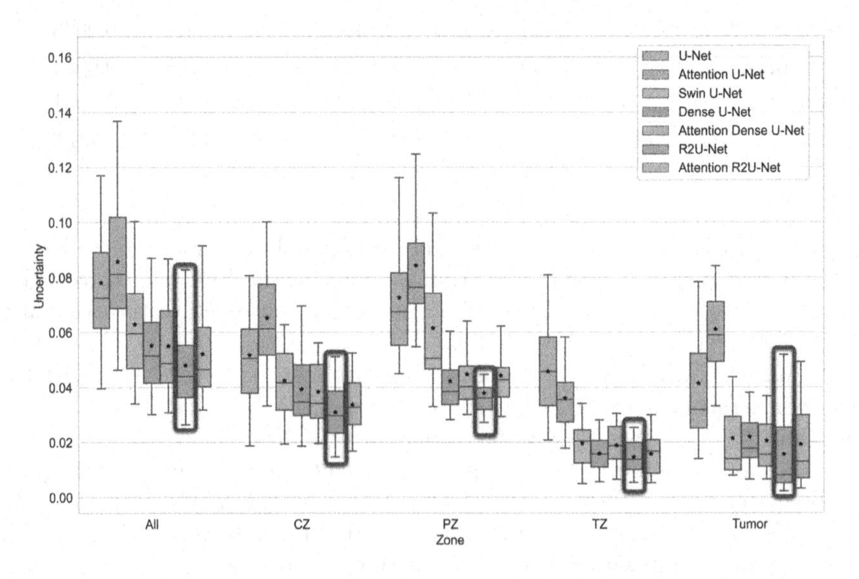

Fig. 1. Comparison of Uncertainty per each class between DL Architectures. The mean uncertainty could be identify with a black star inside each box, and the line represents the median uncertainty value obtained, the best model is indicated with a red box for each zone. (Color figure online)

to segment, models like R2U-Net and Attention R2U-Net managed to achieved a great segmentation performance and uncertainty values in average of those zones in the test set. It is important to notice that both results are correlated. These models managed to be adequately trained to perform the most accurate segmentation task among the others, which can give more confidence to radiologists when using a prostate segmentation tool based in this trained model.

4.2 Qualitative Results

In Fig. 2, a qualitative comparison is presented among the predictions of each model using four different example images from the dataset. The comparison involves all possible combinations of zones. The first two columns display the original T2-MRI image of the prostate and its corresponding ground truth mask. Subsequently, each column represents the average of probabilities obtained from 50 predictions for each model. It can be observed that the first two zone combinations (Image A and B in Fig. 2) are relatively easier for most models, as they produce segmentation that closely resemble the ground truth. However, certain models such as U-Net and Swin U-Net appear to misclassify pixels as TZ even when they are not present in the ground truth. Nevertheless, based on the examples in the test set, the models have been trained effectively to achieve satisfactory segmentation performance on images containing CZ and PZ, and some including TZ.

Regarding the other two combinations that include the tumor, they posed the most complex segmentation challenge with notable variation among models. In Image C of Fig. 2, models like U-Net and Attention Dense U-Net incorrectly classified a TZ region that was not identified in the ground truth. Meanwhile, other models tended to excessively smooth the original segmentation, yielding a

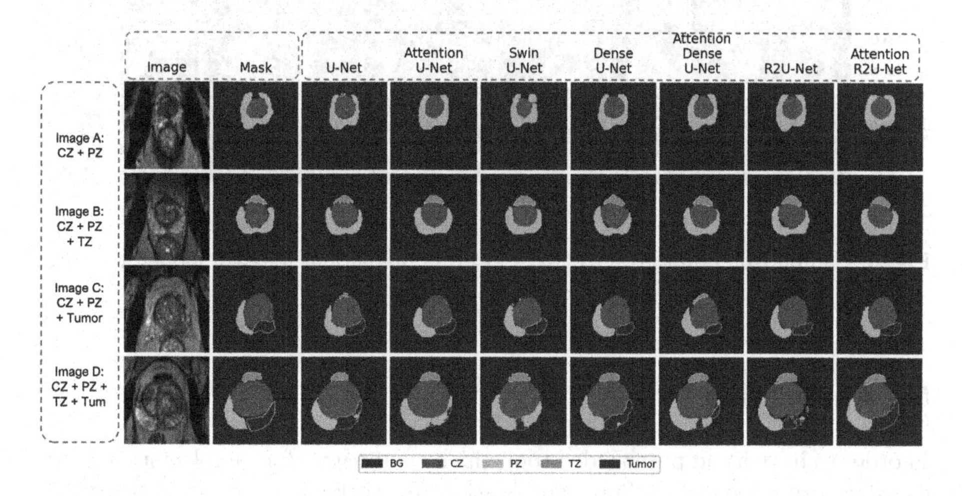

Fig. 2. Comparison of average segmentation after 50 predictions for each model in all the combinations of zones in the dataset.

seemingly good but possibly inaccurate result. However, when visually compared to the ground truth, the best segmentation in this example was achieved by R2U-Net and Attention R2U-Net.

For the last example, most models struggled to accurately segment the tumor. Surprisingly, U-Net and Dense U-Net produced acceptable results, but Attention R2U-Net demonstrated the best overall performance.

Figure 3 illustrates the significance of uncertainty by displaying the same four examples as in the previous figure, along with corresponding uncertainty maps represented as heat maps for each trained model. The temperature of the image indicates the level of uncertainty, with higher temperatures indicating greater uncertainty in those pixels, while lower temperatures indicate higher certainty in the model's pixel segmentation.

The model with the highest uncertainty, particularly around the boundaries of TZ and tumor, is U-Net, followed by Attention U-Net. This observation is evident. Furthermore, as previously mentioned, the first two examples were easier for the models, resulting in relatively low uncertainty across most of them. When dealing with images containing tumors, the inclusion of dense blocks enhanced model certainty. However, the utilization of recurrent residual blocks and attention modules surpassed other models, achieving acceptable predictions in the test set with low uncertainty values, even in TZ and tumor tissues.

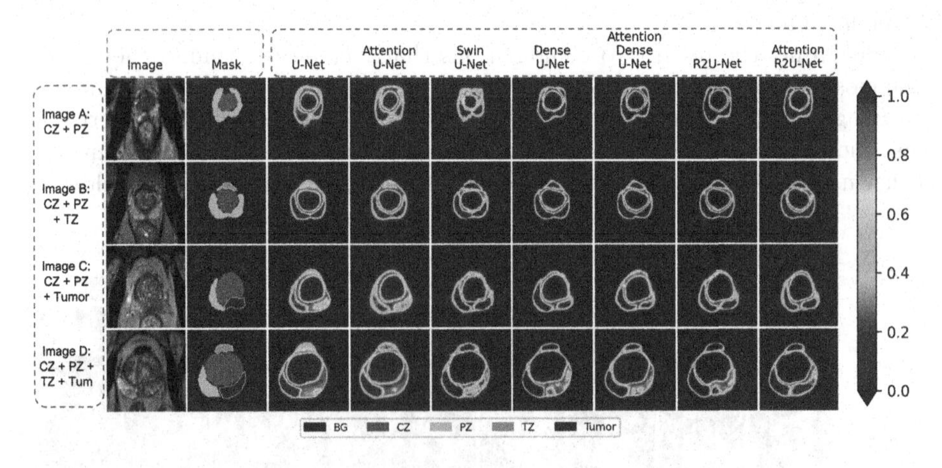

Fig. 3. Comparison of uncertainty maps after 50 predictions for each model with previous examples.

5 Application

In order to have a computer-aided tool which can be used for radiologists or clinicians, we proposed a Web App using Flask framework which we called *'ProstAI'*, and it was designed to have easier access to predict images using the best trained model with MC dropouts: Attention R2U-Net. This app predicts the segmentation mask, as well as the uncertainty map, which is very helpful to indicate the experts which are the pixels where the model has higher uncertainty about their segmentation, an example is shown in Fig. 4.

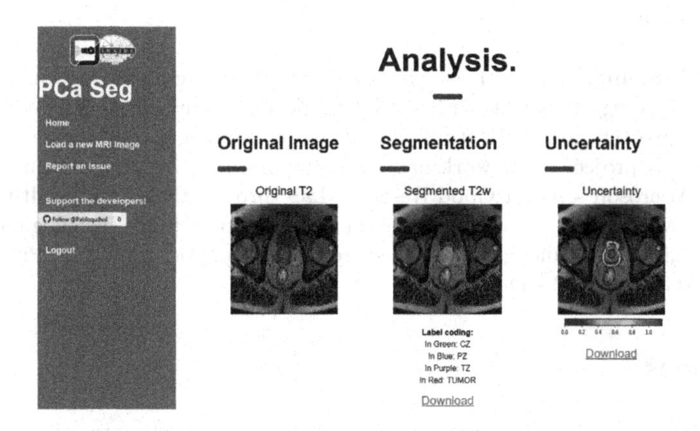

Fig. 4. Example of the analysis page of the *'ProstAI'* app using a prostate image from the Test set.

This tool is proposed for experimental usage, further information about the app and an example of usage can be found in: https://github.com/pabloquihui/ProstAI.

6 Conclusion

This study makes a valuable contribution to prostate cancer segmentation by introducing the segmentation of transition and tumor zones, along with the quantification of uncertainty, which has received limited attention in existing literature. The utilization of a private dataset validated by multiple experts, including two radiologists and two oncologists, enhances the reliability and accuracy of the findings. A comparison of seven different deep learning models was conducted using segmentation metrics, uncertainty scores, and visual inspection. Among these models, Attention R2U-Net emerged as the top-performing approach in both analyses. The inclusion of recurrent residual blocks in U-Net (R2U-Net) notably enhanced the segmentation results by capturing additional contextual information. Furthermore, Attention R2U-Net demonstrated exceptional proficiency in segmenting all prostate zones, exhibiting superior performance in metrics and yielding lower average uncertainty estimated using the MC method. This highlights the positive impact of attention modules on improving segmentation and, more significantly, reducing uncertainty in predictions by focusing on the ROI.

Moreover, a web app has been developed with a focus on experimental use for radiologists. This app provides more accurate, consistent, and faster results and displays the uncertainty map for each predicted image. The uncertainty map provides a visual representation of the pixels in which the model is uncertain about the segmentation, giving radiologists a better idea of the areas that require further analysis.

Acknowledgments. The authors wish to acknowledge the Mexican Council for Science and Technology (CONACYT) for the support in terms of postgraduate scholarships in this project, and the Data Science Hub at Tecnologico de Monterrey for their support on this project. This work has been supported by Azure Sponsorship credits granted by Microsoft's AI for Good Research Lab through the AI for Health program. The authors would also like to thank the financial support from Tecnologico de Monterrey through the "Challenge-Based Research Funding Program 2022". Project ID # E120 - EIC-GI06 - B-T3 - D.

References

1. Sung, H., et al.: Global cancer statistics 2020: globocan estimates of incidence and mortality worldwide for 36 cancers in 185 countries. CA Cancer J. Clin. **71**(3), 209–249 (2021)
2. AstraZeneca. A personalized approach in prostate cancer (2020). https://www.astrazeneca.com/our-therapy-areas/oncology/prostate-cancer.html. Accessed 17 Oct 2021
3. Chen, M., et al.: Prostate cancer detection: comparison of t2-weighted imaging, diffusion-weighted imaging, proton magnetic resonance spectroscopic imaging, and the three techniques combined. Acta Radiologica **49**(5), 602–610 (2008)
4. Haralick, R., Shapiro, L.: Image segmentation techniques. Comput. Vision Graph. Image Process. **29**(1), 100–132 (1985)
5. Aldoj, N., Biavati, F., Michallek, F., Stober, S., Dewey, M.: Automatic prostate and prostate zones segmentation of magnetic resonance images using densenet-like u-net. Sci. Rep. **10**, 08 (2020)
6. Ronneberger, O., Fischer, P., Brox, T.: U-net: convolutional networks for biomedical image segmentation. In: Navab, N., Hornegger, J., Wells, W.M., Frangi, A.F. (eds.) MICCAI 2015. LNCS, vol. 9351, pp. 234–241. Springer, Cham (2015). https://doi.org/10.1007/978-3-319-24574-4_28
7. Cao, H., et al.: Swin-unet: unet-like pure transformer for medical image segmentation. In: Karlinsky, L., Michaeli, T., Nishino, K. (eds.) Computer Vision - ECCV 2022 Workshops, pp. 205–218 (2023). Springer, Cham (2022). https://doi.org/10.1007/978-3-031-25066-8_9
8. Alom, M.Z., Hasan, M., Yakopcic, C., Taha, T.M., Asari, V.K.: Recurrent residual convolutional neural network based on u-net (r2u-net) for medical image segmentation. CoRR, abs/1802.06955 (2018)
9. Basar, S., Ali, M., Ochoa-Ruiz, G., Zareei, M., Waheed, A., Adnan, A.: Unsupervised color image segmentation: a case of rgb histogram based k-means clustering initialization. PLOS ONE **15**(10), 1–21 (2020)
10. Liu, Y., et al.: Exploring uncertainty measures in bayesian deep attentive neural networks for prostate zonal segmentation. IEEE Access **8**, 151817–151828 (2020)

11. Zhu, Q., Du, B., Turkbey, B., Choyke, P.L., Yan, P.: Deeply-supervised CNN for prostate segmentation. CoRR, abs/1703.07523 (2017)
12. Clark, T., Wong, A., Haider, M., Khalvati, F.: Fully deep convolutional neural networks for segmentation of the prostate gland in diffusion-weighted mr images, pp. 97–104 (2017)
13. Oktay, O., et al.: Attention u-net: learning where to look for the pancreas (2018)
14. Li, S., Dong, M., Du, G., Mu, X.: Attention dense-u-net for automatic breast mass segmentation in digital mammogram. IEEE Access **7**, 59037–59047 (2019)
15. Joy, T.T., Sedai, S., Garnavi, R.: Analyzing epistemic and aleatoric uncertainty for drusen segmentation in optical coherence tomography images (2021)
16. Sedai, S., Antony, B., Mahapatra, D., Garnavi, R.: Joint segmentation and uncertainty visualization of retinal layers in optical coherence tomography images using bayesian deep learning (2018)
17. Mata, C., Munuera, J., Lalande, A., Ochoa-Ruiz, G., Benitez, R.: Medicalseg: a medical gui application for image segmentation management. Algorithms **15**(06), 200 (2022)
18. Gal, Y., Ghahramani, Z.: Dropout as a bayesian approximation: representing model uncertainty in deep learning. In: International Conference on Machine Learning, pp. 1050–1059. PMLR (2016)
19. Wu, Y., Wu, J., Jin, S., Cao, L., Jin, G.: Dense-u-net: dense encoder-decoder network for holographic imaging of 3d particle fields. Opt. Commun. **493**, 126970 (2021)

MoSID: Modality-Specific Information Disentanglement from Multi-parametric MRI for Breast Tumor Segmentation

Jiadong Zhang[1], Qianqian Chen[1,2], Luping Zhou[3], Zhiming Cui[1], Fei Gao[1], Zhenhui Li[4], Qianjin Feng[2], and Dinggang Shen[1,5,6](✉)

[1] School of Biomedical Engineering, ShanghaiTech University, Shanghai 201210, China
dgshen@shanghaitech.edu.cn

[2] School of Biomedical Engineering, Southern Medical University, Guangzhou 510515, Guangdong, China

[3] School of Electrical and Information Engineering, The University of Sydney, Sydney, NSW 2006, Australia

[4] Department of Radiology, The Third Affiliated Hospital of Kunming Medical University, Kunming 650118, China

[5] Shanghai United Imaging Intelligence Co., Ltd., Shanghai 200230, China

[6] Shanghai Clinical Research and Trial Center, Shanghai 200052, China

Abstract. Breast cancer is a major health issue, causing millions of deaths each year worldwide. Magnetic Resonance Imaging (MRI) is an effective tool for detecting and diagnosing breast tumors, with various MRI sequences providing comprehensive information on tumor morphology. However, existing methods for segmenting tumors from multi-parametric MRI have limitations, including the lack of considering inter-modality relationships and exploring task-informative modalities. To address these limitations, we propose the Modality-Specific Information Disentanglement (MoSID) framework, which extracts both intra- and inter-modality attention maps as prior knowledge to guide tumor segmentation from multi-parametric MRI. This is achieved by disentangling modality-specific information that provides complementary clues to the segmentation task and generating modality-specific attention maps in a synthesis manner. The modality-specific attention maps are further used to guide modality selection and inter-modality evaluation. Experiment results on a large breast dataset show that the MoSID achieves superior performance over other state-of-the-art multi-modality segmentation methods, and works reasonably well even with missing modalities.

Keywords: Breast tumor · Segmentation · Disentanglement

1 Introduction

Breast cancer is the most common type of malignant neoplasm affecting women [9,12]. Early detection and timely treatment can greatly increase survival rate.

J. Zhang and Q. Chen–Equal Contribution.

© The Author(s), under exclusive license to Springer Nature Switzerland AG 2023
S. Ali et al. (Eds.): CaPTion 2023, LNCS 14295, pp. 94–104, 2023.
https://doi.org/10.1007/978-3-031-45350-2_8

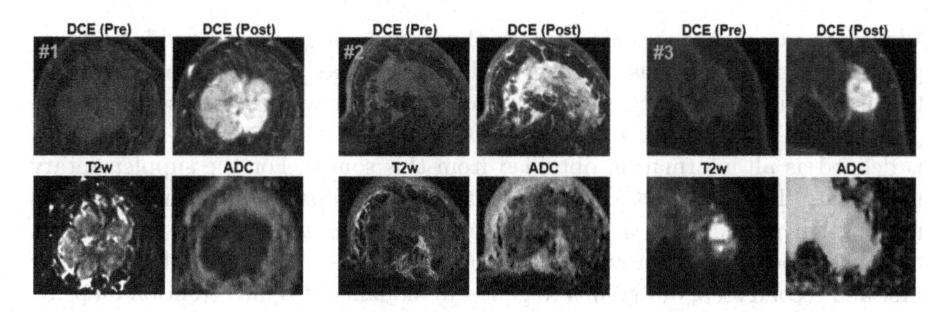

Fig. 1. Multi-parametric breast MR images from three typical cases, including DCE images (pre-contrast and post-contrast), T2w images, and ADC images.

Magnetic resonance imaging (MRI) is an excellent modality for lesion detection and diagnosis, due to its sensitiveness to tumors. In clinics, multiple MRI sequences, e.g., dynamic-contrast enhanced MRI (DCE), T2-weighted image (T2w), and apparent diffusion coefficient (ADC), provide comprehensive information to detect and diagnose tumors. In Fig. 1, compared to the DCE image without contrast agent injection (i.e., pre-contrast image), the tumor region is enhanced and exhibits a higher intensity distribution after contrast agent injection (i.e., post-contrast image). Nonetheless, tumor segmentation from multiple MRI sequences is still challenging. In most cases, tumor regions show higher signal in T2w images and lower signal in ADC images (1st case in Fig. 1). However, some tumors (especially those that are large and have a triple negative molecular type) may contain slime or calcification, resulting in unusual signal intensity distribution in T2w and ADC images (2nd and 3rd cases in Fig. 1). Therefore, these characteristics pose a great challenge for accurately and automatically segmenting breast tumors from multi-parametric MRI.

Many multi-modality segmentation methods typically focus on extracting and combining task-specific information from each modality, but usually overlook the inter-relationships between different modalities. This could limit their ability to fully exploit the complementary information available across multiple modalities and may lead to suboptimal segmentation results. Recently, several disentanglement-based methods have been proposed to improve segmentation performance, by extracting the complementary and additional information of individual modality, i.e., modality-specific information. However, the disentanglement is usually conducted in the feature space through an attention module and lacks of interpretability. Moreover, all of these methods fail to address the questions [5]: (1) Are all modalities for each patient informative for the segmentation task? (2) Will the extracted modality-specific information benefit or damage the segmentation task?

To address these questions, we propose the **Mo**dality-**S**pecific **I**nformation **D**isentanglement (**MoSID**) framework for tumor segmentation from multiple MRI sequences. Based on clinical diagnosis experience, we use the DCE image as the main modality, and T2w and ADC images as supplementary modalities.

We aim to extract the specific and complementary information provided by supplementary modalities based on the main modality as prior knowledge in order to further guide modality selection and evaluation with better segmentation performance. For the main modality (DCE), the modality-specific information is defined as all information obtained from this source. For the supplementary modalities (T2w and ADC), the modality-specific information is the additional information provided by a supplementary modality, which is not shared with the main modality and provides complementary clues about the subject. In the first and second step, our goal is to generate modality-specific attention maps by disentangling task-oriented, modality-specific information from a perspective of information theory. Specifically, we synthesize fake T2w and ADC images by fully leveraging the information from DCE image, representing the shared information between T2w/ADC and DCE. Together with their real T2w/ADC counterparts and DCE, four input cases are sent through a segmentation network to generate intermediate segmentation masks. By comparing the segmentation results of the four cases, we could obtain the modality-specific attention maps. The well-generated modality-specific attention maps are then used to guide modality selection (question 1) and inter-modality voxel-wise evaluation (question 2) in the third step, with modality trusty gating and modality-specific attention modules. Experimental results demonstrate that the disentangled modality-specific attention maps play a crucial role, and obtain significant improvements compared to other state-of-the-art (SOTA) methods. Furthermore, our proposed MoSID framework also performs well in the absence of T2w or ADC modality, indicating framework's great robustness.

2 Methodology

The proposed MoSID framework focuses on utilizing DCE images, which have high resolution and better tissue contrast, as the main modality for tumor segmentation. ADC and T2w images are used as supplementary modalities to extract modality-specific attention to guide the segmentation process. As shown in Fig. 2, the MoSID framework consists of three steps, each of which will be described in detail below.

2.1 Step 1: Image Synthesis

The first step of our proposed MoSID method involves synthesizing fake ADC and T2w images from the DCE images. The aim of this step is to provide the same task-related information as the DCE images, but from a different modality view. To accomplish this, we design two synthesis networks, G_1 and G_2, to map the DCE images to ADC and T2w spaces, respectively. The synthesized ADC images $\hat{x}_{ADC}$ and T2w images $\hat{x}_{T2w}$ retain the structural information from the real DCE images x_{DCE}. The information in the images can be represented using $I(\cdot)$. Thus, we have $I(\hat{x}_{ADC}) = I(G_1(x_{DCE})) \leq I(x_{DCE})$ for the synthesized ADC images, indicating that they contain nearly the same information as the DCE images. The same holds true for the T2w images, with

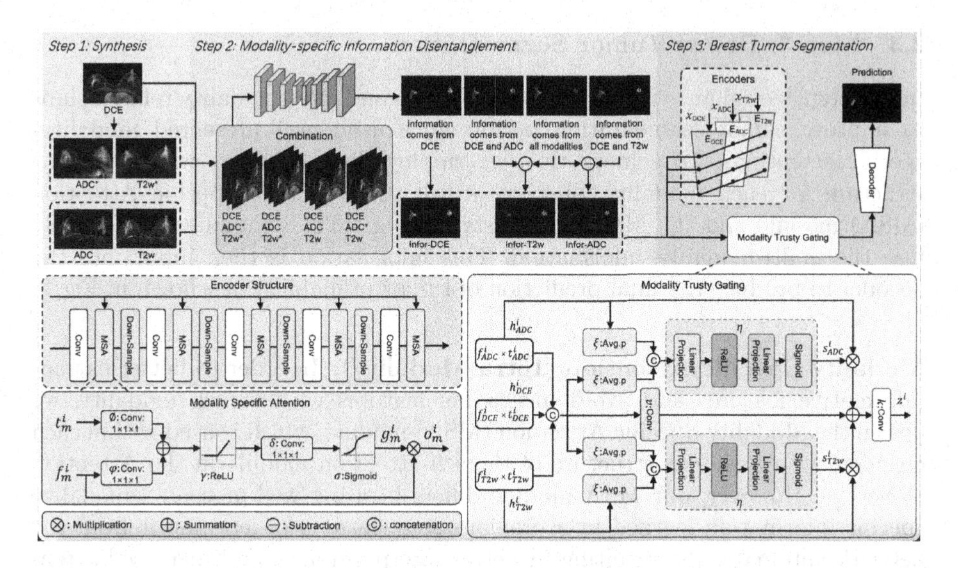

Fig. 2. An overview of the proposed MoSID framework, including image synthesis, modality-specific information disentanglement and breast tumor segmentation. The detailed structure of modality specific attention and modality trusty gating modules are also illustrated.

$I(\hat{x}_{T2w}) = I(G_1(x_{DCE})) \leq I(x_{DCE})$. By comparing the real images (i.e., x_{ADC} and x_{T2w}) with the synthesized images (i.e., $\hat{x}_{ADC}$ and $\hat{x}_{T2w}$), we can easily observe the agreement and differences among the three modalities at the image level.

2.2 Step 2: Modality-Specific Information Disentanglement

In this step, our goal is to find the modality-specific information by comparing synthesized images with real images. We design an information extractor, G_3, to extract task-oriented attention for the downstream segmentation task. Initially, we input the real images $\{x_{DCE}, x_{ADC}, x_{T2w}\}$ into the network and obtain the segmentation probability p_1, which contains information from all modalities. Next, we input different combinations of real and synthesized images into the network and obtain segmentation probability maps p_2, p_3, and p_4. For example, when we input $\{x_{DCE}, \hat{x}_{ADC}, x_{T2w}\}$ into the network, the task-oriented information of p_2 comes from x_{DCE} and x_{T2w} because $\hat{x}_{ADC}$ is synthesized from x_{DCE}. The same also holds for p_3 with input $\{x_{DCE}, x_{ADC}, \hat{x}_{T2w}\}$ (information from x_{DCE} and x_{ADC}) and p_4 with input $\{x_{DCE}, \hat{x}_{ADC}, \hat{x}_{T2w}\}$ (information from x_{DCE}). By subtracting the probability maps, we calculate the specific information from T2w (T_{T2w}) and ADC (T_{ADC}) that differs from the main modality DCE: $T_{T2w} = p_1 - p_3$ and $T_{ADC} = p_1 - p_2$. Since DCE is the main modality with the most important information, we can obtain modality information from the DCE modality with $T_{DCE} = p_4$. The specific information from each modality provides modality trustworthiness and inter-modality attention, which can then be used to guide the downstream tumor segmentation.

2.3 Step 3: Breast Tumor Segmentation

In this step, we enhance the exploration of intra- and inter-modality relationships to improve tumor segmentation performance using well-presented modality-specific attentions. To achieve this, we employ three encoders with the same structure for each modality, and we utilize the Modality Specific Attention (MSA) module and the Modality Trusty Gating (MTG) module to effectively fuse the multi-modality information. This information is then input into the decoder to produce the final prediction of tumor probability, as shown in Fig. 2.

Modality Specific Attention (Intra-Modality). In order to better extract informative features and avoid misleading features within each modality, we design the Modality Specific Attention (MSA) module, which is used within each encoder, with a similar structure of the self-attention module in the Attention U-Net [8]. However, our MSA module differs in using well-presented modality specific information learned in a previous step, instead of learning it from the network automatically, resulting in better interpretability and increased attention on uncertain regions.

Given the original images $x_m \in \mathbb{R}^{1 \times L \times W \times H}$ and the learned modality-specific information $T_m \in \mathbb{R}^{1 \times L \times W \times H}$, where $m \in [DCE, T2w, ADC]$ denotes the modality and L, W, H denote length, width, and height of the input patches, we first calculate multiple feature maps $f_m^i \in \mathbb{R}^{C \times \frac{L}{2^{i-1}} \times \frac{W}{2^{i-1}} \times \frac{H}{2^{i-1}}}$ at different down-sampled scales within each encoder. Here, $i \in [1, 2, 3, 4, 5]$ and C represents the channel number of each feature map. Next, we down-sample the information T_m to obtain multiple maps $t_m^i \in \mathbb{R}^{C \times \frac{L}{2^{i-1}} \times \frac{W}{2^{i-1}} \times \frac{H}{2^{i-1}}}$ at different scales, and then use the MSA module to calculate the correlation attention maps $g_m^i \in \mathbb{R}^{C \times \frac{L}{2^{i-1}} \times \frac{W}{2^{i-1}} \times \frac{H}{2^{i-1}}}$ using the equation:

$$g_m^i = \sigma(\gamma(\delta(\phi t_m^i + \varphi f_m^i + b_\phi + b_\varphi) + b_\delta), \tag{1}$$

where ϕ, φ, δ are convolution kernel weights, as shown in Fig. 2, and $b_\phi, b_\varphi, b_\delta$ are corresponding biases. γ and σ are ReLU and Sigmoid activation functions, respectively. The correlation attention maps more accurately capture the modality-specific information for tumor region localization compared to t_m^i, and we can obtain localized features $o_m^i \in \mathbb{R}^{C \times \frac{L}{2^{i-1}} \times \frac{W}{2^{i-1}} \times \frac{H}{2^{i-1}}}$ by calculating $o_m^i = f_m^i \times g_m^i$ as MSA output.

Modality Trusty Gating (Inter-Modality). As discussed previously, some modalities have abnormal signal intensities that can affect the segmentation accuracy. To overcome this challenge, we design a Modality Trusty Gating (MTG) module that could distinguish task-related information by preserving positive features and removing negative features. The MTG module calculates the confidence scores between the main modality (DCE) and supplementary modalities (T2w and ADC) to evaluate the contribution of each modality to the segmentation task.

To calculate the confidence scores, we need the feature maps from the encoder f_m^i and the down-sampled modality-specific attention map t_m^i. The confidence score for the T2w modality (${s^i}_{T2w}$) is calculated as follows:

$$ {s^i}_{T2w} = \eta(\xi(h^i_{T2w}) :: \xi(h^i_{T2w} :: h^i_{DCE} :: h^i_{ADC})), \quad (2) $$

where ξ represents average pooling, :: represents feature concatenation, and η represents a series of operations including linear projection, ReLU, and Sigmoid activation functions, as illustrated in Fig. 2. Similarly, the confidence score for the ADC modality (${s^i}_{ADC}$) is calculated as:

$$ {s^i}_{ADC} = \eta(\xi(h^i_{ADC}) :: \xi(h^i_{T2w} :: h^i_{DCE} :: h^i_{ADC})). \quad (3) $$

These confidence scores are then used to combine the supplementary modality features with the main modality features, using the following equation:

$$ z^i = \kappa(\upsilon(h^i_{T2w} :: h^i_{DCE} :: h^i_{ADC}) + {s^i}_{T2w} \times {h^i}_{T2w} + {s^i}_{ADC} \times {h^i}_{ADC}), \quad (4) $$

where κ and υ are two convolution operators. The final multi-modality features fed into the decoder have trusty information and produce a final robust segmentation prediction.

3 Experiments and Results

3.1 Dataset and Implementation Details

In this study, a large dataset consisting of 415 cases collected from 413 patients are used. Each case includes two DCE images (pre- and post-contrast image), one T2w image, and one ADC image. To reduce the impact of patient motion during the scan, both T2w and ADC images are well-aligned with DCE images using the ANTs algorithm [1]. Experienced radiologists manually annotate breast tumors in the DCE images, which served as the ground truth. In the data preprocessing stage, all images are resampled to a common space with a voxel size of $1\times1\times1\,\text{mm}^3$ and normalized to the range of $[0, 1]$ using min-max normalization. The data are randomly split into 249 cases for training (60%), 83 cases for validation (20%), and the remaining 83 cases for testing (20%). To accommodate GPU memory limitations, large-scale 3D patches with a voxel size of $128\times128\times96$ are extracted for training. Random data augmentation techniques such as flipping and rotation are applied to reduce the risk of overfitting. To evaluate the tumor segmentation performance, three commonly used metrics were adopted, including Dice similarity coefficient (DSC, 0%–100%), Sensitivity (SEN, 0%–100%), and Average surface distance (ASD, *mm*).

In our proposed framework, we have two synthesis networks (for synthesizing DCE to T2w and DCE to ADC), a modality-specific information extraction network, and a segmentation network. The two synthesis networks and the information extraction network are based on RU-Net, which includes four down/up-sampling operations. The experiments are conducted on PyTorch platform using

Table 1. Breast tumor segmentation performance in terms of DSC, SEN and ASD.

Method	Publication	DSC (%) ↑	SEN (%) ↑	ASD (mm) ↓
Robust-Mseg [2]	MICCAI'19	77.70 ± 4.57	75.12 ± 4.91	7.10 ± 4.72
Cross-Model [7]	MICCAI'19	79.72 ± 4.08	76.46 ± 4.77	5.27 ± 4.01
CSAD [11]	JBHI'21	72.92 ± 3.97	72.99 ± 5.05	8.89 ± 3.02
RFNet [3]	ICCV'21	74.02 ± 5.19	68.87 ± 5.62	6.14 ± 4.22
MAML [14]	MICCAI'21	80.34 ± 3.42	83.03 ± 3.57	6.36 ± 2.89
MMEF-UNet [6]	MICCAI'22	79.89 ± 3.23	79.78 ± 3.73	6.79 ± 2.75
NestedFormer [10]	MICCAI'22	79.95 ± 3.49	83.62 ± 3.62	7.44 ± 2.79
mmFormer [13]	MICCAI'22	79.90 ± 3.20	**83.64 ± 2.89**	7.26 ± 2.90
MoSID (Ours)	CaPTion'23	**83.98 ± 2.52**	82.92 ± 3.41	**2.66 ± 1.38**

two NVIDIA TESLA V100 GPUs with 32GB of memory. The initial learning rate for the synthesis networks is set at 0.0002 and decreases by half every 50 epochs. For the segmentation network, we adopt a RU-Net model with three encoders for each modality as the baseline. The initial learning rate is set at 0.005 and decreased by half every 50 epochs. More details can be viewed in the released code repository.

3.2 Segmentation Performance Analysis

Comparison with SOTA Multi-modal Segmentation Methods. We have compared our proposed MoSID framework with several SOTA methods for multi-modal tumor segmentation, including Robust-Mseg [2], Cross-Model [7], CSAD [11], RFNet [3], MAML [14], MMEF-UNet [6], NestedFormer [10] and mmFormer [13]. It is worth noting that Cross-Model is specifically developed for breast tumor segmentation, while MAML is designed for liver tumor segmentation, and CSAD is developed for prostate tumor segmentation. The other methods are all developed for multi-modal brain tumor segmentation.

As shown in Table 1, the proposed MoSID method surpasses the performance of other methods. In comparison to latent feature fusion techniques, such as [2,3,7,11], MoSID demonstrates a better ability to distinguish between trustworthy and uncertain regions in each modality, which results in improved segmentation performance. Specifically, MoSID achieves 4.2% increase in DSC and 2.6 mm reduction in ASD compared to the Cross-Model method. The proposed MoSID method also offers superior performance compared to both late fusion methods (such as MAML and MMEF-UNet) and transformer-based fusion methods (such as mmFormer and NestedFormer). While late fusion methods use statistical convolution kernels to fuse modalities, they do not take into account the differences between patients. In contrast, MoSID dynamically adjusts modality fusion based on global-local modality trust, using learned modality-specific information attention with MSA and MTG modules. As a result, it achieves better

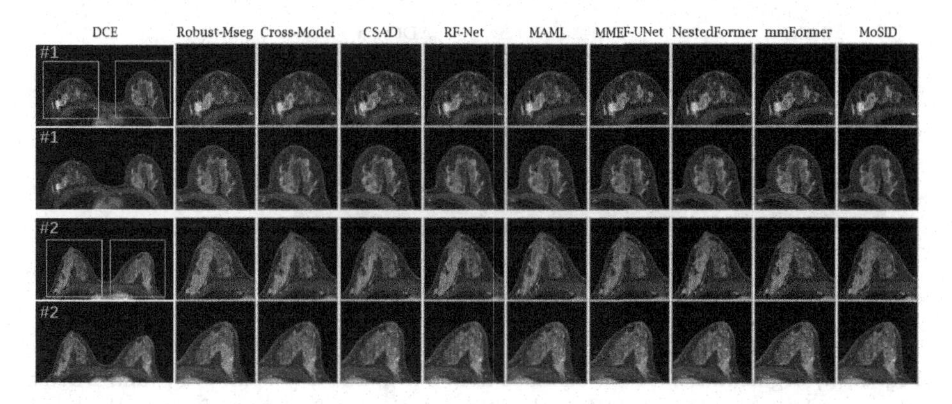

Fig. 3. Visual comparison of tumor segmentation results by different methods on two typical cases. For each case, both left and right breast are illustrated. The red contours are the ground-truth annotations, and the green contours are the predictions from each methods. (Color figure online)

segmentation performance. Specifically, compared to MAML, MoSID results in 3.6% increase in DSC and 3.7 mm decrease in ASD. Similarly, compared with transformer-based methods, MoSID is not susceptible to misleading information in certain modalities of certain patients, leading to overall better segmentation performance.

The visualization results further support the superiority of MoSID. As seen in Fig. 3, each column displays the original DCE-MRI and the corresponding segmentation results (highlighted by orange and blue boxes) for each method. The red contour represents the ground-truth annotation and the green contour shows the segmentation result from each method, which are overlapped for better visualization. It is evident that MoSID produces segmentation results that are more in line with the ground-truth annotation than the other methods. Additionally, in regions where the DCE-MRI displays misleading enhancement, MoSID produces correct results without over-segmentation, potentially resulting in significant improvement for ASD metric, demonstrating its improved performance.

In our method, we separate the shared and specific information of each modality for fine-grained segmentation, similar to uncertainty-based segmentation methods that use Gaussian approximations to detect uncertain regions. For instance, the Dropout Bayesian Network (DBNet [4]) approximates Gaussian processes using different types of dropout during inference. However, DBNet still falls short in effectively exploring the complex modality information, which may result in misleading information and unsatisfactory performance. Figure 4(a) overlays the probability map from our method and DBNet for each modality. For DCE images, our method identifies high uncertainty values on the right breast, whereas DBNet only contains a few regions with lower uncertainty. Furthermore, DBNet does not provide modality-specific information for T2w and ADC

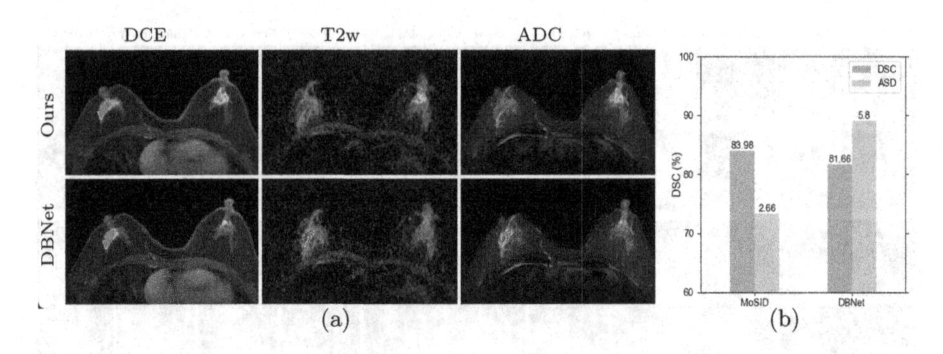

Fig. 4. (a) Segmentation attention maps from our methods (the first row) and DBNet (the second row) for each modality. (b) Segmentation performance of our method and DBNet in terms of DSC and ASD.

Table 2. Segmentation performance of ablation study in terms of DSC, SEN and ASD.

Baseline	MSA	MTG	DSC (%) ↑	SEN (%) ↑	ASD (mm) ↓
✓			76.96 ± 4.74	75.45 ± 5.16	8.72 ± 7.15
✓	✓		82.06 ± 3.65	82.91 ± 3.96	3.28 ± 1.57
✓		✓	81.49 ± 3.16	82.10 ± 3.69	3.95 ± 1.94
✓	✓	✓	**83.98 ± 2.52**	**82.92 ± 3.41**	**2.66 ± 1.38**

images, which are only distributed on small boundary regions. In contrast, our method separates the modality-specific information for T2w and ADC images as attention maps for improved segmentation, resulting in better performance, as shown in Fig. 4(b).

Ablation Studies. To further evaluate the impact of each component in the framework, we conduct a series of ablation studies, as summarized in Table 2. The framework's performance is improved by the specially designed Modality Spatial Attention (MSA) module, which effectively leverages modality-specific spatial information, resulting in 5.1% improvement in DSC. The Modality Trusty Gating (MTG) module removes misleading information in a global manner, enabling the network to dynamically adapt to each case. This results in a significant improvement in segmentation performance, such as 4.5% increase in DSC. Our full method, MoSID (Baseline + MSA + MTG), achieves the best results, with 7.0%, 7.5%, and 6.1 mm improvement in DSC, SEN, and ASD metrics, respectively.

Missing Modality Segmentation. We would like to emphasize that missing modality scenario for breast cancer is relatively rare, as DCE scans are required for clinical diagnosis and ADC/T2w scans only take a few additional minutes to collect. Nevertheless, we also demonstrate the capability of MoSID for missing modality segmentation, which can be applied to other multi-modality segmentation tasks, such as brain tumor segmentation. We use synthesized images to replace real images for breast tumor segmentation.

When only the DCE modality is available, the DSC and ASD are 70.68% and 7.57 mm respectively. On the other hand, when the ADC (T2w) modality is missing, the DSC and ASD are 81.42% (74.11%) and 4.77 mm (4.18 mm) respectively. These results highlight the importance of using multiple modalities for breast tumor segmentation. Additionally, our MoSID is robust in handling missing modality scenario and achieves satisfactory performance.

4 Conclusion

The paper introduces a new approach for breast tumor segmentation in multi-parametric MRI scans, called as the MoSID framework. This framework focuses on disentangling task-specific information for each imaging modality, which is then utilized to guide the segmentation process with the help of modality specific attention and modality trusty gating modules. The results of the experiments conducted on a comprehensive dataset of multi-parametric MRI scans indicate that the MoSID framework significantly outperforms state-of-the-art multi-modality segmentation methods. The effective use of the disentangled modality-specific attention maps play a crucial role and can be extended to other tumor segmentation tasks.

Acknowledgments. This work was supported in part by The Key R&D Program of Guangdong Province, China (grant number 2021B0101420006), National Natural Science Foundation of China (grant number 62131015), and Science and Technology Commission of Shanghai Municipality (STCSM) (grant number 21010502600).

References

1. Avants, B.B., Tustison, N.J., Song, G., Cook, P.A., Klein, A., Gee, J.C.: A reproducible evaluation of ants similarity metric performance in brain image registration. Neuroimage **54**(3), 2033–2044 (2011)
2. Chen, C., Dou, Q., Jin, Y., Chen, H., Qin, J., Heng, P.-A.: Robust multimodal brain tumor segmentation via feature disentanglement and gated fusion. In: Shen, D., et al. (eds.) MICCAI 2019. LNCS, vol. 11766, pp. 447–456. Springer, Cham (2019). https://doi.org/10.1007/978-3-030-32248-9_50
3. Ding, Y., Yu, X., Yang, Y.: Rfnet: region-aware fusion network for incomplete multi-modal brain tumor segmentation. In: Proceedings of the IEEE/CVF International Conference on Computer Vision, pp. 3975–3984 (2021)
4. Gal, Y., Ghahramani, Z.: Dropout as a bayesian approximation: representing model uncertainty in deep learning. In: International Conference on Machine Learning, pp. 1050–1059. PMLR (2016)
5. Han, Z., Yang, F., Huang, J., Zhang, C., Yao, J.: Multimodal dynamics: dynamical fusion for trustworthy multimodal classification. In: Proceedings of the IEEE/CVF Conference on Computer Vision and Pattern Recognition, pp. 20707–20717 (2022)

6. Huang, L., Denoeux, T., Vera, P., Ruan, S.: Evidence fusion with contextual discounting for multi-modality medical image segmentation. In: International Conference on Medical Image Computing and Computer-Assisted Intervention, pp. 401–411. Springer, Heidelberg (2022). https://doi.org/10.1007/978-3-031-16443-9_39
7. Li, C., Sun, H., Liu, Z., Wang, M., Zheng, H., Wang, S.: Learning cross-modal deep representations for multi-modal MR image segmentation. In: Shen, D., et al. (eds.) MICCAI 2019. LNCS, vol. 11765, pp. 57–65. Springer, Cham (2019). https://doi.org/10.1007/978-3-030-32245-8_7
8. Oktay, O., et al.: Attention u-net: learning where to look for the pancreas. arXiv preprint arXiv:1804.03999 (2018)
9. Venturelli, S., Leischner, C., Helling, T., Renner, O., Burkard, M., Marongiu, L.: Minerals and cancer: overview of the possible diagnostic value. Cancers **14**(5), 1256 (2022)
10. Xing, Z., Yu, L., Wan, L., Han, T., Zhu, L.: Nestedformer: nested modality-aware transformer for brain tumor segmentation. In: International Conference on Medical Image Computing and Computer-Assisted Intervention, pp. 140–150. Springer, Heidelberg (2022). https://doi.org/10.1007/978-3-031-16443-9_14
11. Zhang, G., et al.: Cross-modal prostate cancer segmentation via self-attention distillation. IEEE J. Biomed. Health Inf. **26**, 5298–5309 (2021)
12. Zhang, J., et al.: A robust and efficient AI assistant for breast tumor segmentation from DCE-MRI via a spatial-temporal framework. Patterns **4**(9), 1–14 (2023)
13. Zhang, Y., et al.: mmformer: multimodal medical transformer for incomplete multimodal learning of brain tumor segmentation. arXiv preprint arXiv:2206.02425 (2022)
14. Zhang, Y., et al.: Modality-aware mutual learning for multi-modal medical image segmentation. In: de Bruijne, M., et al. (eds.) MICCAI 2021. LNCS, vol. 12901, pp. 589–599. Springer, Cham (2021). https://doi.org/10.1007/978-3-030-87193-2_56

Cancer/Early cancer Surveillance

Colonoscopy Coverage Revisited: Identifying Scanning Gaps in Real-Time

George Leifman(✉), Idan Kligvasser, Roman Goldenberg, Ehud Rivlin, and Michael Elad

Verily Research, Haifa, Israel
gleifman@google.com

Abstract. Colonoscopy is the most widely used medical technique for preventing Colorectal Cancer, by detecting and removing polyps before they become malignant. Recent studies show that around 25% of the existing polyps are routinely missed. While some of these do appear in the endoscopist's field of view, others are missed due to a partial coverage of the colon. The task of detecting and marking unseen regions of the colon has been addressed in recent work, where the common approach is based on dense 3D reconstruction, which proves to be challenging due to lack of 3D ground truth and periods with poor visual content. In this paper we propose a novel and complementary method to detect deficient local coverage in real-time for video segments where a reliable 3D reconstruction is impossible. Our method aims to identify skips along the colon caused by a drifted position of the endoscope during poor visibility time intervals. The proposed solution consists of two phases. During the first, time segments with good visibility of the colon and *gaps* between them are identified. During the second phase, a trained model operates on each *gap*, answering the question: "Do you observe the same scene before and after the gap?" If the answer is negative, the endoscopist is alerted and can be directed to the appropriate area in real-time. The second phase model is trained using a contrastive loss based on the auto-generated examples. Our method evaluation on a dataset of 250 procedures annotated by trained physicians provides sensitivity of 75% with specificity of 90%.

Keywords: Colonoscopy · Coverage · Self-supervised Learning

1 Introduction

Colorectal cancer is one of the most preventable cancers, as early detection and through screening is highly effective. The most common screening procedure is optical colonoscopy – visually examining the surface of the colon for abnormalities such as colorectal lesions and polyps. However, performing a thorough examination of the entire colon surface is proven to be quite challenging due to unavoidable poor visibility segments of the procedure. As a consequence, improperly

© The Author(s), under exclusive license to Springer Nature Switzerland AG 2023
S. Ali et al. (Eds.): CaPTion 2023, LNCS 14295, pp. 107–118, 2023.
https://doi.org/10.1007/978-3-031-45350-2_9

inspected regions may lead to a lower detection rate of polyps. Indeed, recent studies have shown that approximately 25% of polyps are routinely missed during a typical colonoscopy procedure [16].

Various efforts to automatically detect and mark non-inspected regions of the colon are reported in recent publications, where the common approach relies on the creation of a dense 3D reconstruction of the colon's shape [8,11,24,26,27,33]. However, such a reconstruction based on video solely is a challenging task, and especially so in colonoscopy, in which reflections, low-texture content, frequent changes in lighting conditions and erratic motion are common. As a consequence, while the above 3D approach has promise, it is limited to segments of the video exhibiting good visual quality.

In this work we propose a novel real-time approach for detecting deficient local coverage, complementing the 3D reconstruction methods mentioned above. Our proposed strategy provides a reliable, stable and robust solution for the grand challenge posed by temporal periods of poor visual content, such as camera blur, poor camera positioning, occlusions due to dirt and spayed water, and more. The proposed method consists of two main phases. During the first, we identify time segments with good visibility of the colon and gaps of poor visibility between them. For this purpose we train a binary classifier, leveraging a small set of annotated images and a self-supervised training scheme. During the second phase, we train an ML model that aims to answer the following question for each gap: *Do you observe different scenes before and after the gap?* (see Fig. 1). If the answer is positive, we suspect a loss of coverage due to an unintentional drift of the endoscope position, and therefore alert the endoscopist accordingly in real-time to revisit the area.

The second phase model is designed to generate low-dimensional frame-based descriptors that are used for scene-change detection via a simple Cosine distance evaluation. This network is trained using a contrastive loss based on automatically generated positive and negative pairs of video segments. These training examples are sampled from good-visibility segments of real colonoscopy videos, where the translational speed of the endoscope can be reliably estimated.

To evaluate our method we introduce a dataset of 250 colonoscopy procedures (videos). Two doctors have been asked to evaluate up to 5 gaps per video and decide whether they suspect loss of coverage there. The evaluation of our method using this annotated dataset provides sensitivity of 75% with specificity of 90%.

We note that our task of same-scene detection in the colon is related to image retrieval [2,25,32] and geo-localization [5,19,20]. There is also some similarity to techniques employed for face recognition [1,4,10,18] and person re-identification [7,21,29]. In the narrower domain of colonoscopy, the only closely related work we are aware of is reported in [22]. While their technique for location recognition is related to our scene descriptor generation, their eventual tasks are markedly different, and so are the evaluation protocols. Nevertheless, for completeness of this work, we evaluate our scene descriptors on their dataset and show that our method outperforms their results.

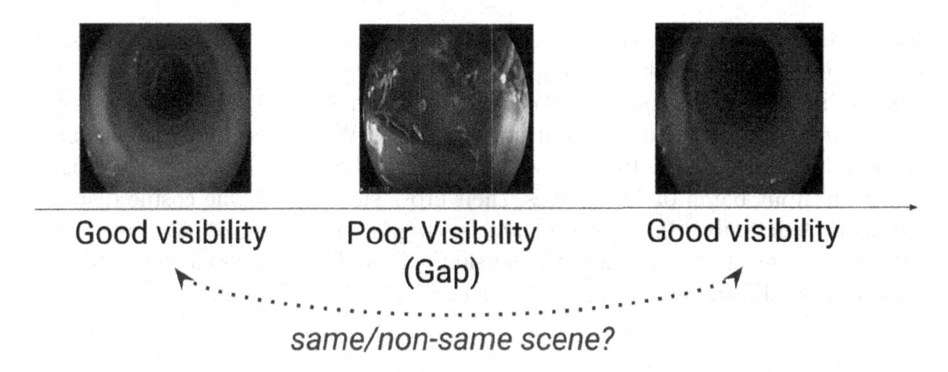

Fig. 1. Our solution starts by detecting time segments with good visibility of the colon and gaps between them. For each such gap we answer the question: *Do you observe different scenes before and after the gap?* If the answer is positive, the endoscopist is alerted to revisit the area in real-time.

To summarize, this work offers three main contributions:

- We present a novel stable, robust and accurate method for detecting deficient local coverage in real-time for periods with poor visual content.
- Our coverage solution complements the 3D reconstruction approach, covering cases beyond its reach;
- We introduce a novel self-supervised method for generating frame-based descriptors for scene change-detection in colonoscopy videos.

This paper is organized as follow: Sect. 2 describes Phase I of our method, aiming to identify time segments with good visibility of the colon and gaps between them. Phase II of our method is presented in Sect. 3, addressing the *same-scene* question by metric learning. Section 4 summarizes the results of our experiments and Sect. 5 concludes the paper.

2 Method: Phase I – Visibility Classification

Our starting point is a frame-based classification of the visibility content. We characterize good visibility frames as those having a clear view of the tubular structure of the colon. In contrast, poor visibility frames may include severe occlusions due to dirt or sprayed water, a poor positioning of the camera - being dragged on the colon walls, or simply blurred content due to rapid motion.

In order to solve this classification task, we gather training and validation annotated datasets by experts. Operating on 85 different colonoscopy videos, 5 good visibility segments and 5 poor ones were identified in each. A naive supervised learning of a classifier leads to an unsatisfactory 84% accuracy on the validation set due to insufficient data. In an attempt to improve this result, we adopt a semi-supervised approach. First, we pre-trained an encoder on large (1e6) randomly sampled frames using simCLR [9]. This unsupervised learning embeds the frames such that similar ones (obtained by augmentations of the same frame) are close-by, while different frames (the rest of the frames in the batch) are pushed away. Given the learned encoder, we train a binary classifier on the resulting embeddings using the labeled data. Since the dimension of the embedding vectors is much smaller than the original frame sizes (512 vs. 224^2), this approach leads to far better accuracy of 93%. We further improve the above by smoothing the predictions based on their embeddings, as shown in Fig. 2. For each input batch of 512 frames, their cross-similarities (the cosine distance between their embedding vectors) are leveraged, such that similar frames are also encouraged to be assigned to the same class. This improves the per-frame accuracy on the validation set up to 94%.

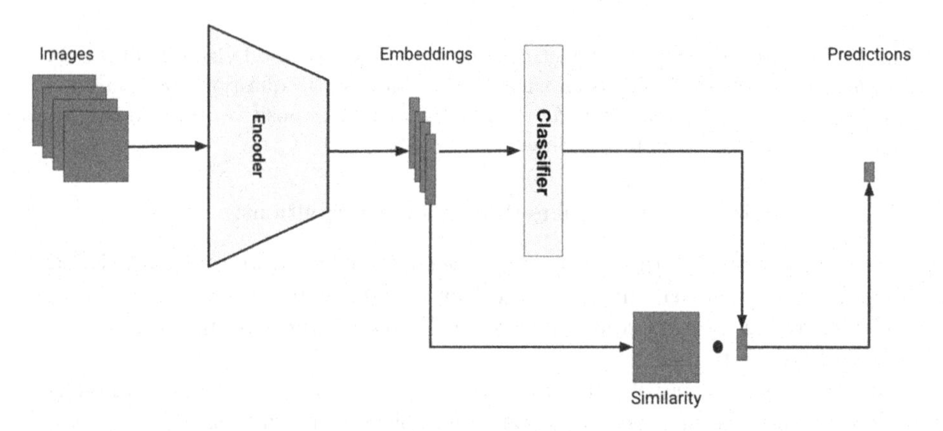

Fig. 2. To achieve high accuracy visibility classifier, we train an encoder in an unsupervised manner and then train a binary classifier the resulting embeddings using the labeled data. Further improvement is made by smoothing predictions based on similarity distances, resulting in 94% accuracy on the validation set.

To conclude, the trained classifier provides a partitioning of the time axis into disjoint intervals of good or poor visibility. In order to further relax these intervals, we apply a median filter with window size of 10 frames.

3 Method: Phase II – Gaps with Loss of Coverage

After partitioning the procedure timeline into periods with good visibility and gaps between them, our goal now is to identify gaps with a potential loss of coverage, defined as exhibiting a change of the scene between their ends. In order to compare scenes before and after a gap, we learn distinctive frame descriptors. These vectors are compared via a simple distance measure for addressing the same/not-same scene question. While the direct approach towards this task would be to gather a training set of many thousands of such gaps along with their human annotation, we introduce a much cheaper, faster, and easier alternative based on a self-supervised approach. In this section we describe all these ingredients in greater detail.

3.1 Scene Descriptors

Assume that a training set of the form $\{F_1^k, F_2^k, c_k\}_{k=1}^N$ is given to us, where F_1^k and F_2^k are two frames on both sides of a given gap, and c_k is their label, being $c_k = 1$ for the same scene and 0 otherwise. N is the size of this training data, set in this work to be $N = 1e5$ examples. We design a neural network $f = T_\Theta(F)$ that embeds the frame F to the low-dimensional vector $f \in \mathbb{R}^{512}$, while accommodating our desire to serve the same/not-same scene task. More specifically, our goal is to push same-scene descriptor-pairs to be close-by while forcing pairs of different scenes to be distant, being the essence of contrastive learning, which has been drawing increased attention recently [3,9,13,30]. Therefore, we train $T_\Theta(\cdot)$ to minimize the loss function

$$L(\Theta) = \sum_{k=1}^{N}(2c_k - 1)d\left(T_\Theta(F_1^k), T_\Theta(F_2^k)\right) \tag{1}$$
$$= \sum_{\{c_k=1\}_k} d\left(T_\Theta(F_1^k), T_\Theta(F_2^k)\right) - \sum_{\{c_k=0\}_k} d\left(T_\Theta(F_1^k), T_\Theta(F_2^k)\right).$$

In the above expression, $d(\cdot,\cdot)$ stands for a distance measure. In this work we use the Cosine similarity $d(f_1, f_2) = 1 - f_1^T f_2/\|f_1\|_2\|f_2\|_2$.

Creating the Training Data: Constructing the training set $\{F_1^k, F_2^k, c_k\}_{k=1}^N$ might be a daunting challenge if annotations by experts are to be practiced. We introduce a fully automatic alternative that builds on a reliable displacement estimation of the endoscope, accessible in good visibility video segments of any real colonoscopy. This displacement can be evaluated by estimating the optical-flow between consecutive frames (see [23,28]) and estimating the amount of flow trough the frame boundary [15] (see Fig. 3).

Given any time interval of good visibility content, the cumulative directional transnational motion can be estimated rather accurately. Thus, starting with such a video segments, and randomly marking an inner part of it of a random

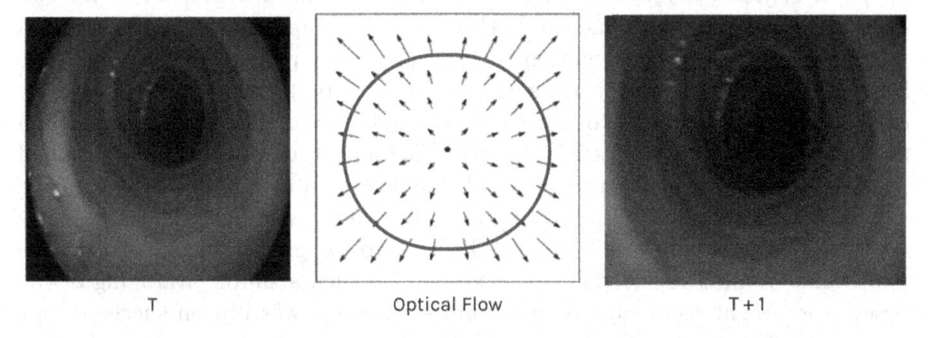

Fig. 3. Endoscope displacement estimation is based on optical-flow calculation between consecutive frames using the amount of flow trough the frame boundary (see [15]).

length of 5–30 s as a pseudo-gap, we can define frames on both its ends as having the same scene or not based on the accumulated displacement. Figure 4 presents the whole process of creating training examples this way, easily obtaining triplets $\{F_1^k, F_2^k, c_k\}$.

Our attempts to improve the above contrastive training scheme by introducing a margin, as practiced in [12] and employing a "soft-max" loss version [30], did not bring a significant improvement. A technique that delivered a benefit is to pre-train the network T_Θ in fully unsupervised way using simCLR [9] (as in Sect. 2), and proceed along the above contrastive learning scheme.

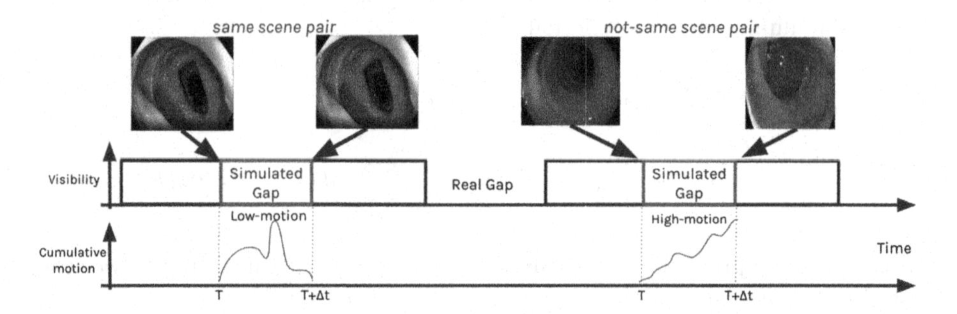

Fig. 4. We simulate random artificial gaps of various duration in good-visibility video segments, estimate the endoscope motion within these simulated gaps, and get this way reliable training examples for our overall task. Gaps associated with low accumulated motion contribute a 'same-scene' training example ($c_k = 1$), while high-motion gaps refer to a different scene pair ($c_k = 0$).

3.2 Gap Classification

With a simple machinery of a distance evaluation of the frame descriptors on both ends of any gap, we are now equipped to answer our main questions: *Is there a potential loss of coverage during this poor-visibility video segment? Has the probe drifted away form its original position?* As this distance evaluation can be applied over various frames on both sides of the gap, the challenge is to find a reliable fusion of the information within these many pairs of frames. While we have experimented with various such options, the best results are achieved by calculating a single descriptor for the scenes before and after the gap, and then comparing these using a Cosine distance. This unified descriptor, $\bar{f}$, is obtained by a weighted average of the individual descriptors in a segment of 2 s on each side, f_i, as follows: $\bar{f} = \sum_i f_i w_i / \sum_i w_i$, where $w_i = v_i e^{-s_i}$, v_i and s_i are the raw visibility score and the temporal distance to the gap, both referring to the i-th frame. While the effectiveness of employing such a simple averaging of the descriptors might seem surprising, a similar strategy was proven successful for face recognition from multiple views in [31].

4 Results

As explained in Sect. 3, first we generate per-frame scene descriptors and then employ them to detect the gaps with potential loss of coverage. This section starts from presenting the evaluation of the stand-alone scene descriptors and compares them to SOTA. Then we describe the dataset of the annotated gaps and present the evaluation of our gap classifier on this dataset.

4.1 Scene Descriptors

We evaluate our scene descriptors on the recently released dataset for colonoscopic image retrieval – Colon10K [22]. This dataset contains 20 short sequences (10,126 images), where the positive matching images were manually labeled and verified by an endoscopist. We follow the setup and the evaluation metrics described in [22]. In total, they have 620 retrieval tasks (denoted by "*all*"), while 309 tasks use the intervals that are not direct neighbor frames of their queries as positives (denoted by "*indirect*"). We use the data from Colon10K for the evaluation purposes only. Table 1 compares the results to those reported in [22]. *Rank-1 recognition rate* is the percentage of tasks in which the most similar to the query image is true positive. The *Mean average precision* is the area under the precision-recall curve. For both metrics our method outperforms [22] for both "*all*" and "*indirect*" tasks.

Table 1. Comparison of our scene descriptor generation to [22] on the Colon10K dataset. In all the evaluated metrics our method outperforms [22].

	Rank-1 recognition rate		Mean average precision (mAP)	
Method	*all*	*indirect*	*all*	*indirect*
[22]	0.9032	0.8058	0.9042	0.8245
Our	0.9131	0.8173	0.9723	0.9112

4.2 Gap Classification

Figure 5 demonstrates an example of our gap classification. In the top row the case with no loss of coverage is presented, where the scene before and after the gap is the same. In the bottom row the case with potential loss of coverage is presented, where the scenes before and after the gap are different.

For quantitative evaluation of our gap classification we introduce a dataset of 250 colonoscopy procedures (videos) from five different hospitals. We have automatically identified between 2 to 5 *true* gaps in each video and presented these to doctors for their annotation – whether a loss of coverage is suspected. Each gap was evaluated by two doctors and the ones without a consensus ($\sim$25%) were omitted. This resulted with 750 gaps having high-confidence annotations,

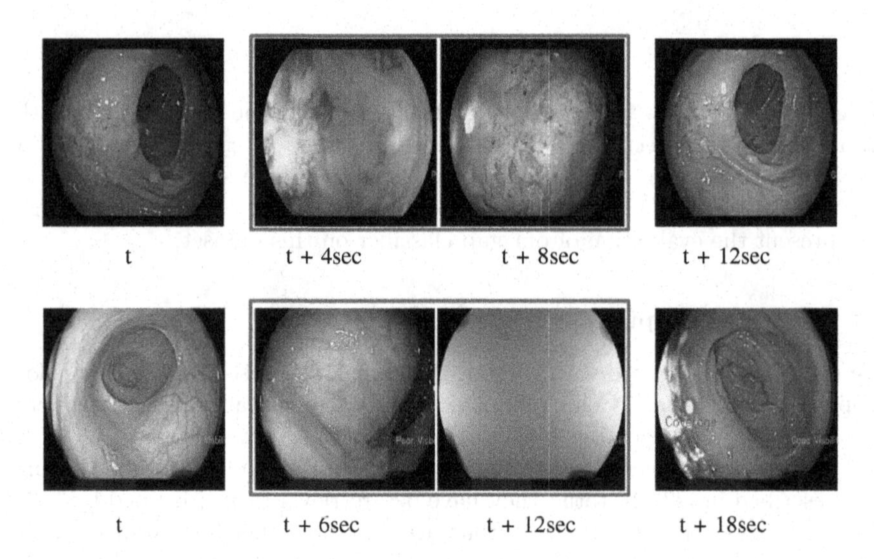

Fig. 5. Qualitative evaluation of our gap classification method. Top: no loss of coverage detected (the scene before and after the gap is the same). Bottom: potential loss of coverage detected (the scenes before and after the gap are different). The gaps with the poor visibility frames are highlighted in red. (Color figure online)

150 of which are marked as exhibiting a loss of coverage. Figure 6 presents the ROC of our direct gap classification method evaluated on the whole dataset of 750 gaps. At the working point of 10% false alarms (alert on gaps with no coverage loss) we cover 75% of gaps with real coverage loss. The area under curve (AUC) is 0.9, which usually indicates a high-accuracy classifier.

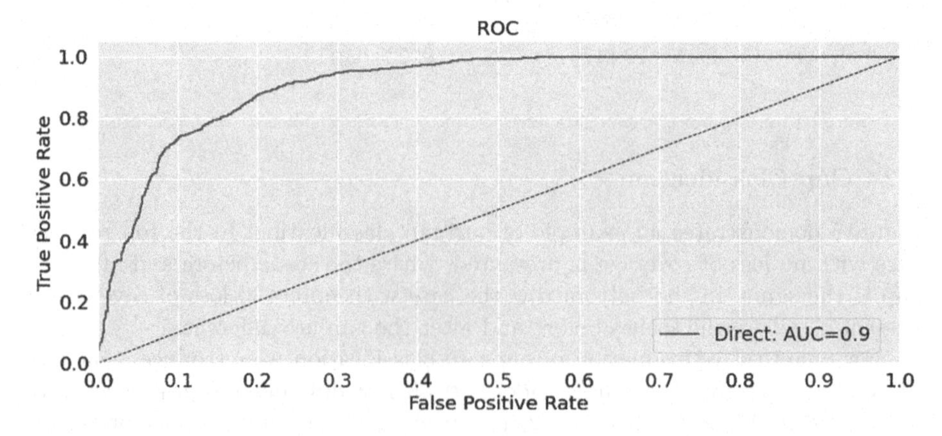

Fig. 6. Direct gap classification: ROC curve evaluated on the whole dataset (750 gaps).

The above classification exploits the information before and after the gap, while completely disregarding the information about the gap itself. Having a dataset of annotated true gaps, we can improve this accuracy by a *supervised* learning that exploits the gap characteristics. We thus split the dataset of the annotated gaps 50:50 to training and evaluation. Since we have a very limited amount of the training examples we use a low-dimensional classifier – Gradient Boosting [6] – that operates on the following input data: (i) A 32-bin histogram of the similarity matrix' values between frames two seconds before and after the gap; (ii) A 32-bin histogram of the visibility scores two seconds before and after the gap; (iii) A 32-bin histogram of the visibility scores inside the gap; and (iv) The duration of the gap. We performed class-balancing using up-sampling with augmentations before training.

Table 2 compares the original approach to the supervised one, summarizing the contribution of different input features to the final accuracy (measured by AUC). In the supervised approach we use one half of the dataset for training, thus the evaluation is performed using the other half of the dataset for both the original and the supervised approaches. In the first approach we also explored a classification based on the gap duration only, getting an AUC of 0.651, being higher than random but lower than employing frame similarities. Weighing the scene descriptors by the visibility scores (see Sect. 3) improves the AUC by 2%. In the supervised approach both gap duration and visibility scores inside the gap provide a substantial contribution of 2% each to the AUC.

Table 2. Impact of various features on the AUC, evaluated on 375 test gaps.

Method	Features				
	Frame similarities	Gap duration	Visibility inside the gap	Visibility outside the gap	AUC
		✓			0.651
Original	✓				0.876
	✓			✓	0.896
	✓				0.881
Supervised	✓	✓			0.898
	✓	✓	✓		0.929
	✓	✓	✓	✓	0.932

4.3 Implementation Details

We employ MobilenetV3 [14] backbone both for the visibility classification and for the scene descriptors. The inference time of each model for a single input image of 256×256 is below 20 ms, which makes our approach real-time. The descriptor network was pre-trained for 200 epochs using simCLR [9] with learning rate 0.005 and then we continued training on the auto-generated training data for additional 100 epochs with learning rate 0.001, using Adam optimizer [17].

5 Conclusion

This work presents a novel method for the detection of deficient local coverage in real-time for periods with poor visual content, complementing any 3D reconstruction alternative for coverage assessment of the colon. Our method starts with an identification of time segments with good visibility of the colon and *gaps* between them. For each such *gap* we train an ML model that tests whether the scene has changed during the gap, alerting the endoscopist in such cases to revisit a given area in real-time. Our learning constructs frame-based descriptors for same scene detection, leveraging a self-supervised approach for generating the required training set. For the evaluation of the gap classification results we have built a dataset of 250 colonoscopy videos with annotations of gaps with deficient local coverage.

We plan to extend our approach to a guidance of the endoscopist to the exact place where the coverage was lost, and use our scene descriptors for bookmarking points of interest in the colon. In order to do it we propose to use exhaustive descriptor similarity comparison in a predefined time interval. This exhaustive search can also address the limitation of our approach during the movement of the endoscope backwards.

References

1. Adjabi, I., Ouahabi, A., Benzaoui, A., Taleb-Ahmed, A.: Past, present, and future of face recognition: a review. Electronics **9**(8), 1188 (2020)
2. Ali, S., Rittscher, J.: Efficient video indexing for monitoring disease activity and progression in the upper gastrointestinal tract. In: 2019 IEEE 16th International Symposium on Biomedical Imaging (ISBI 2019), pp. 91–95. IEEE (2019)
3. Bachman, P., Hjelm, R.D., Buchwalter, W.: Learning representations by maximizing mutual information across views. Adv. Neural Inf. Process. Syst. **32**, 1–11 (2019)
4. Bae, G., et al.: Digiface-1m: 1 million digital face images for face recognition. In: Proceedings of the IEEE/CVF Winter Conference on Applications of Computer Vision, pp. 3526–3535 (2023)
5. Berton, G., Masone, C., Paolicelli, V., Caputo, B.: Viewpoint invariant dense matching for visual geolocalization. In: Proceedings of the IEEE/CVF International Conference on Computer Vision, pp. 12169–12178 (2021)
6. Brownlee, J.: XGBoost With python: gradient boosted trees with XGBoost and scikit-learn. In: Machine Learning Mastery (2016)
7. Chen, H., Wang, Y., Lagadec, B., Dantcheva, A., Bremond, F.: Joint generative and contrastive learning for unsupervised person re-identification. In: Proceedings of the IEEE/CVF Conference on Computer Vision and Pattern Recognition, pp. 2004–2013 (2021)
8. Chen, R.J., Bobrow, T.L., Athey, T., Mahmood, F., Durr, N.J.: Slam endoscopy enhanced by adversarial depth prediction. arXiv preprint arXiv:1907.00283 (2019)
9. Chen, T., Kornblith, S., Norouzi, M., Hinton, G.: A simple framework for contrastive learning of visual representations. In: International Conference on Machine Learning, pp. 1597–1607. PMLR (2020)

10. Deng, J., Guo, J., Xue, N., Zafeiriou, S.: Arcface: additive angular margin loss for deep face recognition. In: Proceedings of the IEEE/CVF Conference on Computer Vision and Pattern Recognition. pp. 4690–4699 (2019)
11. Freedman, D., et al.: Detecting deficient coverage in colonoscopies. IEEE Trans. Med. Imaging **39**(11), 3451–3462 (2020)
12. Hadsell, R., Chopra, S., LeCun, Y.: Dimensionality reduction by learning an invariant mapping. In: 2006 IEEE Computer Society Conference on Computer Vision and Pattern Recognition (CVPR 2006), vol. 2, pp. 1735–1742. IEEE (2006)
13. He, K., Fan, H., Wu, Y., Xie, S., Girshick, R.: Momentum contrast for unsupervised visual representation learning. In: Proceedings of the IEEE/CVF Conference on Computer Vision and Pattern Recognition, pp. 9729–9738 (2020)
14. Howard, A., et al.: Searching for mobilenetv3. In: Proceedings of the IEEE/CVF International Conference on Computer Vision, pp. 1314–1324 (2019)
15. Kelner, O., Weinstein, O., Rivlin, E., Goldenberg, R.: Motion-based weak supervision for video parsing with application to colonoscopy. In: Proceedings of the "What is Motion for?" Workshop, ECCV (2022)
16. Kim, N.H., et al.: Miss rate of colorectal neoplastic polyps and risk factors for missed polyps in consecutive colonoscopies. Intest. Res. **15**(3), 411 (2017)
17. Kingma, D.P., Ba, J.: Adam: a method for stochastic optimization. arXiv preprint arXiv:1412.6980 (2014)
18. Kortli, Y., Jridi, M., Al Falou, A., Atri, M.: Face recognition systems: a survey. Sensors **20**(2), 342 (2020)
19. Lin, T.Y., Belongie, S., Hays, J.: Cross-view image geolocalization. In: Proceedings of the IEEE Conference on Computer Vision and Pattern Recognition, pp. 891–898 (2013)
20. Lin, T.Y., Cui, Y., Belongie, S., Hays, J.: Learning deep representations for ground-to-aerial geolocalization. In: Proceedings of the IEEE Conference on Computer Vision and Pattern Recognition, pp. 5007–5015 (2015)
21. Lin, Y., Xie, L., Wu, Y., Yan, C., Tian, Q.: Unsupervised person re-identification via softened similarity learning. In: Proceedings of the IEEE/CVF Conference on Computer Vision and Pattern Recognition, pp. 3390–3399 (2020)
22. Ma, R., et al.: Colon10k: a benchmark for place recognition in colonoscopy. In: 2021 IEEE 18th International Symposium on Biomedical Imaging (ISBI), pp. 1279–1283. IEEE (2021)
23. Oliveira, M., Araujo, H., Figueiredo, I.N., Pinto, L., Curto, E., Perdigoto, L.: Registration of consecutive frames from wireless capsule endoscopy for 3d motion estimation. IEEE Access **9**, 119533–119545 (2021)
24. Posner, E., Zholkover, A., Frank, N., Bouhnik, M.: C3 fusion: consistent contrastive colon fusion, towards deep slam in colonoscopy. arXiv preprint arXiv:2206.01961 (2022)
25. Radenović, F., Tolias, G., Chum, O.: Fine-tuning CNN image retrieval with no human annotation. IEEE Trans. Pattern Anal. Mach. Intell. **41**(7), 1655–1668 (2018)
26. Rau, A., et al.: Implicit domain adaptation with conditional generative adversarial networks for depth prediction in endoscopy. Int. J. Comput. Assist. Radiol. Surg. **14**(7), 1167–1176 (2019)
27. Shao, S., et al.: Self-supervised monocular depth and ego-motion estimation in endoscopy: appearance flow to the rescue. Med. Image Anal. **77**, 102338 (2022)
28. Sun, D., Yang, X., Liu, M.Y., Kautz, J.: Pwc-net: cnns for optical flow using pyramid, warping, and cost volume. In: Proceedings of the IEEE Conference on Computer Vision and Pattern Recognition, pp. 8934–8943 (2018)

29. Wang, D., Zhang, S.: Unsupervised person re-identification via multi-label classification. In: Proceedings of the IEEE/CVF Conference on Computer Vision and Pattern Recognition, pp. 10981–10990 (2020)
30. Wang, F., Liu, H.: Understanding the behaviour of contrastive loss. In: Proceedings of the IEEE/CVF Conference on Computer Vision and Pattern Recognition, pp. 2495–2504 (2021)
31. Wolf, L., Hassner, T., Taigman, Y.: Descriptor based methods in the wild. In: Workshop on Faces in 'Real-Life' Images: Detection, Alignment, and Recognition (2008)
32. Yan, C., Gong, B., Wei, Y., Gao, Y.: Deep multi-view enhancement hashing for image retrieval. IEEE Trans. Pattern Anal. Mach. Intell. **43**(4), 1445–1451 (2020)
33. Zhang, S., Zhao, L., Huang, S., Ye, M., Hao, Q.: A template-based 3d reconstruction of colon structures and textures from stereo colonoscopic images. IEEE Trans. Med. Rob. Bionics **3**(1), 85–95 (2020)

ColNav: Real-Time Colon Navigation for Colonoscopy

Netanel Frank(✉), Erez Posner, Emmanuelle Muhlethaler, Adi Zholkover, and Moshe Bouhnik

Intuitive Surgical, Inc., 1020 Kifer Road, Sunnyvale, CA, USA
{netanel.frank,erez.posner,emmanuelle.muhlethaler,
adi.zholkover,moshe.bouhnik}@intusurg.com

Abstract. Colorectal cancer screening through colonoscopy continues to be the dominant global standard, as it allows identifying pre-cancerous or adenomatous lesions and provides the ability to remove them during the procedure itself. Nevertheless, failure by the endoscopist to identify such lesions increases the likelihood of lesion progression to subsequent colorectal cancer. Ultimately, colonoscopy remains operator-dependent, and the wide range of quality in colonoscopy examinations among endoscopists is influenced by variations in their technique, training, and diligence. This paper presents a novel real-time navigation guidance system for Optical Colonoscopy (OC). Our proposed system employs a real-time approach that displays both an unfolded representation of the colon and a local indicator directing to un-inspected areas. These visualizations are presented to the physician during the procedure, providing actionable and comprehensible guidance to un-surveyed areas in real-time, while seamlessly integrating into the physician's workflow. Through coverage experimental evaluation, we demonstrated that our system resulted in a higher polyp recall (PR) and high inter-rater reliability with physicians for coverage prediction. These results suggest that our real-time navigation guidance system has the potential to improve the quality and effectiveness of OC and ultimately benefit patient outcomes.

Keywords: Colonoscopy · Coverage · Real-time systems

1 Introduction

Colorectal cancer (CRC) is a significant public health issue, with over 1.9 million new cases diagnosed globally in 2020 [5]. It is one of the most preventable types of cancer [14], and early detection is crucial for preventing its progression [13,15]. The most commonly used screening method is optical colonoscopy (OC) [10], which visually inspects the mucosal surface of the colon for abnormalities such as colorectal lesions. However, the process of detecting CRC in its early stages

N. Frank and E. Posner—These authors contributed equally to this work.

© The Author(s), under exclusive license to Springer Nature Switzerland AG 2023
S. Ali et al. (Eds.): CaPTion 2023, LNCS 14295, pp. 119–131, 2023.
https://doi.org/10.1007/978-3-031-45350-2_10

can be difficult, since performing a comprehensive examination of the colon using OC alone can be challenging, resulting in certain regions of the colon not being fully examined and potentially reducing the rate of polyp detection.

To address this problem, researchers have conducted extensive studies to propose assistive technologies that set out to provide clinicians with a better understanding of the procedure quality. Most existing methods focus on estimating measures of quality, such as the withdrawal time, or on reconstructing a 3D model of the colon from a video sequence of the procedure, some even try automating the entire scan procedure using robotic autonomous colonoscopy navigation [16]. Despite the advancements in technology that allow for the prediction of 3D structures from images, there is still a significant gap in providing useful and actionable information to clinicians during the procedure in real-time. Current methods for detecting un-surveyed regions, which usually show a 3D visualization of the colon, are not designed to be easily understood, or interacted with, during the procedure. They may not align with the camera view; making it difficult for physicians to understand where they need to move the endoscope to survey missing regions. Other measures of quality, such as coverage per frame, or withdrawal time, do not provide clear, usable information to assist during the procedure in capturing un-surveyed regions. In this paper, we present ColNav, a novel real-time solution that (i) utilizes an unfolded representation of the colon to localize the endoscope within the colon, (ii) introduces a local indicator that directs the physician to un-surveyed areas and (iii) is robust to real-life issues such as tracking loss. Our approach estimates the centerline and unfolds the scanned colon from a 3D structure to a 2D image in a way that not only calculates the coverage, but also provides augmented guidance to un-surveyed areas without disrupting the physician's workflow. To the best of our knowledge, this is the first coverage based, real-time guidance system for colonoscopies.

2 Related Work

In recent years, there has been an abundance of papers exploring various aspects of quality measures for colonoscopy, with the goal of assisting clinicians and improving the overall quality of care.

SLAM for colonoscopy approaches usually rely on estimating a 3D reconstruction of the colon and post-processing it in order to estimate the un-surveyed regions (holes). Posner et al. [17] utilized deep features to better track the camera position, presented tracking loss recovery and loop closure capabilities to create a consistent 3D model. Ma et al. [11] reconstructed fragments of the colon using Direct Sparse Odometry (DSO) [7] and a Recurrent Neural Network (RNN) for depth estimation. However their output is not easily understood nor meant to be interacted with during the procedure, making them less likely to be adopted by physicians or impact the clinical outcome.

Direct coverage estimation methods [4,8], aim to predict the coverage on a segment-by-segment basis by estimating what fraction of the colon has been

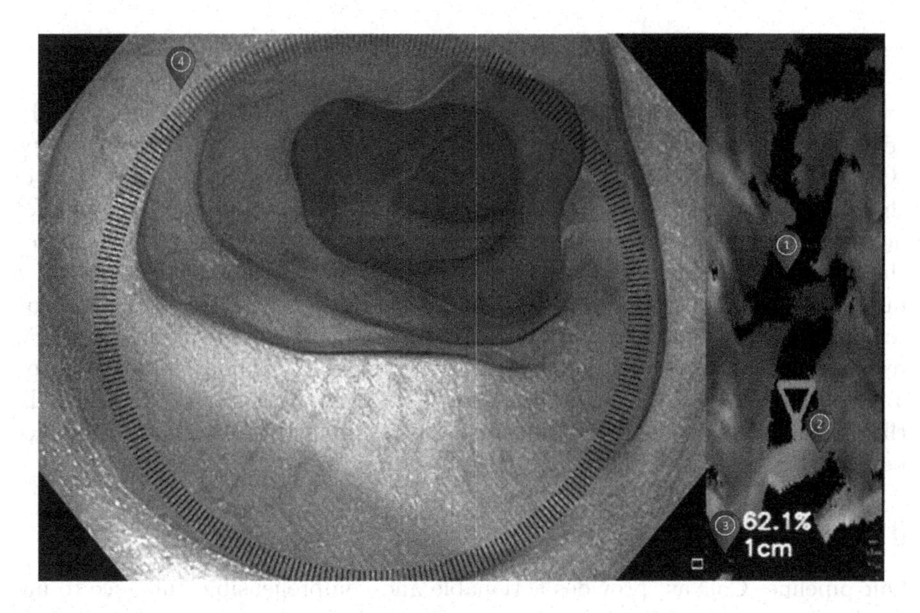

Fig. 1. Our novel, real-time colonoscopy navigation. Flattened image of the colon (right): (1) un-surveyed areas as black pixels, (2) camera location, (3) coverage percentage and length covered. Endoscope view (left): (4) the local compass indicator directing the physician to look up (ticks highlighted in red). (Color figure online)

viewed in any given segment. Freedman et al. [8] used a CNN to perform depth estimation for each frame followed by coverage estimation. As it was trained in a supervised manner using synthetic ground-truth coverage, it cannot be easily generalized to real data. Blau et al. [4] proposed an unsupervised learning technique for detecting deficient colon coverage segments modeled as curved cylinders. However, their method does not run in real-time.

Indirect Quality Objective Measurements: Objective measurements of quality in colonoscopy are important for minimizing subjective biases and variations among endoscopists. Yao et al. [23] proposed an auto-detection of non-Informative frames and Zhou et al. [24] predicted the bowel preparation scores every 30 s during the withdrawal phase of the procedure. However, these techniques lack the ability to provide real-time information to the physician during the procedure.

Virtual colon unfolding is a well known visualisation technique for virtual colonoscopy (VC), where the colon is inspected by analysing the output of a CT scan. In such cases, a 3D mesh of the entire colon can be extracted from the CT image volume and mapped onto a 2D grid, providing the physician with a fast and convenient way to inspect the colon mucosa and find polyps. A number of solutions have been proposed to perform this mapping [9,19,20,22]. In many cases, the solution uses, as an intermediate step, the computation of the *centerline*, a single continuous line spanning the colon. Although most of these methods tend to be computationally expensive, Sudarsky et al. [19] proposed a fast unfolding method based on the straightening of the colon mesh using the centerline. Using colon unfolding to visualize missed areas in optical colonoscopy (OC) has been proposed in a number of works [1–3,11]. A cylindrical model of the colon is used by [1–3] to unwrap and combine frames. This method does not detect hidden areas behind haustral folds, nor does it handles frames in which the camera is oriented towards the colon walls. Furthermore, it does not run in real time and the results were only demonstrated on relatively short colon segments. The method of Ma et al [11] is based on a more accurate SLAM reconstruction, but colon unfolding was done offline, for validation purposes, on a few disconnected colon segments, using a single straight line as centerline. To the best of our knowledge, no prior work shows a map of the entire colon surface being consistently updated during the procedure.

3 Method Overview

Our pipeline, ColNav, provides actionable and comprehensible guidance to unsurveyed areas in real-time and is seamlessly integrated into the physician's workflow. While scanning the colon, the physician is presented with 2 screens as can be seen in Fig. 1. The colon unfolded image, presented on the right, shows the three-dimensional (3D) colon flattened into a two-dimensional (2D) image. Black pixels in the image indicate unseen areas, which were missed during the scan. The location of the endoscope distal tip (camera) in the unfolded colon is visualized as the green camera frustum marker.

When the physician withdraws the endoscope the green marker moves down, while the flattened image gets updated with new rows at the bottom. These rows represent the newly scanned portion of the colon. If the physician decides to move the endoscope back into the colon, the camera marker will move up. The relevant areas which were re-scanned will then be updated, but no new rows will be created in the flattened representation. This enables the physician to know whether to move the endoscope forwards or backwards to reach the missed regions (holes). Coverage percentage and the overall travelled length, computed on the scanned portion of the colon, are displayed as well.

The left screen in Fig. 1 is the main endoscope view with a local compass indicator directing to un-inspected areas (holes). Once the endoscope is near an un-inspected area, the relevant sections of the compass indicator will be highlighted in red, directing the physician towards the areas in need of further examination.

The ColNav algorithm, as depicted in Fig. 2, consists of three major parts: (*i*) centerline estimation, (*ii*) multi-segment 3D to 2D unfolding, and (*iii*) local indicator (navigation compass). The inputs of ColNav are the depth and pose of each RGB frame. To obtain these inputs, we employ our previous work C^{3}Fusion [17] as a SLAM module. C^{3}Fusion was specifically designed to deal with real life colonoscopy issues, filter out blurry/ non-informative frames, and 3D reconstruct the colon from OC.

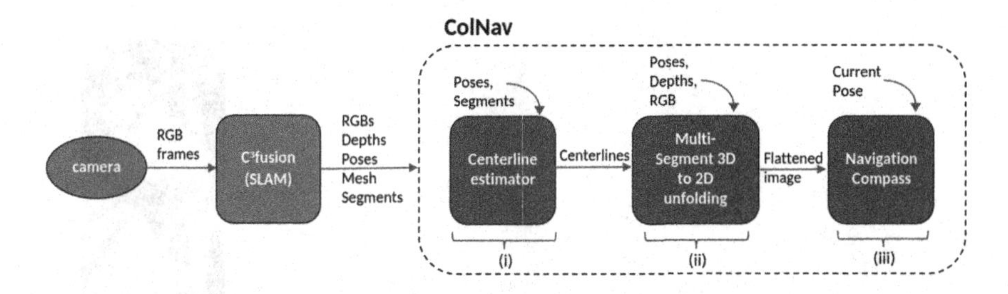

Fig. 2. Block diagram of our ColNav method, comprised of three main parts, (*i*) centerline estimation, (*ii*) multi-segment 3D to 2D unfolding, and (*iii*) local indicator (navigation compass), as well as C^{3}Fusion [17] which is employed as our SLAM module

The first component of ColNav is a robust method for centerline estimation. In the second component, the overall 2D scene representation is obtained by merging the depth, pose, and RGB of all frames into a single flattened representation of the colon. Both components are described in Sect. 3.1. The third component, the creation of the navigation compass, is described in Sect. 3.2. In real-life scenarios, C^{3}Fusion may lose track, resulting in the creation of a new segment when the last frame cannot be connected to any previous frame. Alternatively, loop closure may occur, where two disjoint segments are merged, and their poses are subsequently updated. Our system accommodates these scenarios by (a) adjusting the previous centerline approximation based on updated poses and (b) de-integrating frames that have changed location in the flattened image and re-integrating them with their new pose (see Sect. 3.3). In cases of tracking loss, the flattened image shows separate segments with red lines that can be merged if tracking recovery occurs.

3.1 Centerline and Colon Unfolding

In our proposed solution, the 3D representation of each frame is obtained by back-projecting the depth map into a point cloud. It is then mapped onto a 2D unfolded representation, using an algorithm analogous to that described in [19]. In particular, the centerline is used for straightening the reconstructed colon and dividing it into cross-sections perpendicular to the centerline. Each cross-sectional slice corresponds to a row within the two-dimensional flattened image, see Fig. 3.

The colon centerline, sometimes also referred to as the medial axis, is usually defined as a single connected line, spanning the colon and situated at its center, away from the colon walls [21]. In our case, the centrality requirement is partly relaxed, as we observed that a shift of the centerline away from the center of the colon has little effect on the unfolding. In ColNav, unlike prior works, the centerline and the flattened image are also updated in real time. This creates new requirements for the centerline: (1) Fast computation. (2) Consistency over time

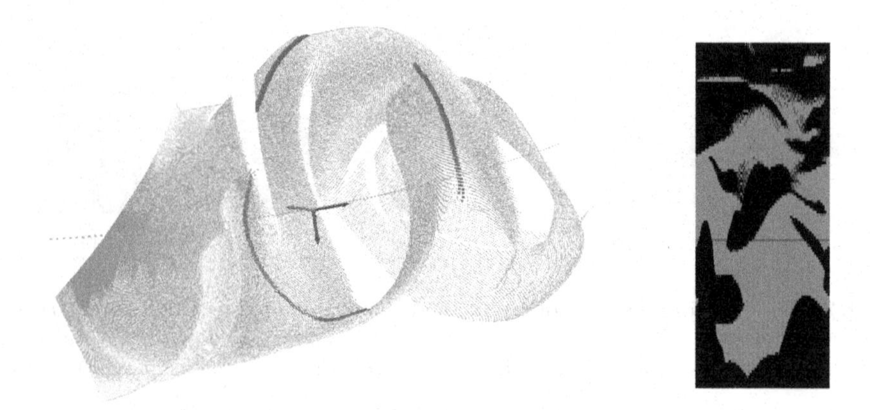

Fig. 3. On the left: 3D point cloud of a single frame (blue), centerline (green), vertices associated with a specific cross-section on the centerline (red) and the camera pose indicated by the 3 axis vectors. On the right: the flattened image with corresponding cross-section (red). Note that the holes in the cross-section match the black pixels in the corresponding row. (Color figure online)

relatively to the camera trajectory. To support these requirements, the centerline is estimated from the camera trajectory poses.

The centerline algorithm contains the following steps: (1) Filtering outlier poses from the trajectory. (2) Constructing or updating a graph G of the trajectory, with camera positions as nodes and edges connecting nodes within a threshold distance. (3) Calculating or updating the shortest path length l between each node $n \in G$. (4) Binning of the trajectory points according to l. (5) Fitting a B-spline [6] to an aggregate of the trajectory points in each bin. Each time the trajectory poses are updated, steps (1)–(5) are computed and a new centerline is re-calculated.

Camera Position Indicator: The camera position for each frame is given by the SLAM module and is noted by $T_i = \{(R_i, t_i) | R_i \in SO(3), t_i \in \mathbb{R}^3\}_{i=1}^{N}$ with N the number of frames in the sequence. To represent the endoscope current location s_e on the centerline of size K, The endoscope position t_i is projected on the centerline $C = \{c_k \in \mathbb{R}^3\}_{k=1}^{K}$ by querying the centerline KDTree.

$$s_e = \arg\min_{c_k \in C} ||t_i - c_k||^2, t_i \in \mathbb{R}^3 \tag{1}$$

3.2 Navigation Compass

The navigation compass serves as a local indicator that visually guides the physician to areas that have been missed. Specifically, the compass ticks are highlighted in red to indicate which specific sections of the colon require further inspection. Based on the camera position along the centerline s_e, the coverage information is extracted from the unfolded image F, where each column represents b_θ - the rotation angle bin around the centerline axis at the endoscope location s_e, with $\theta \in \{0, ..., 2\pi\}$.

When $F(s_e, b_\theta)$ - the corresponding pixel in the extracted row - is black (meaning, it wasn't covered), the navigation compass tick will be highlighted in red, otherwise it will remain dark. To make the navigation compass invariant to camera roll, the camera orientation is projected on the centerline and the relative angle offset is computed to compensate for misalignment between the centerline and the camera pose. Figure 3 depicts the extracted row, selected from the flattened image according to the camera location.

3.3 Unfolding Real-Time Dynamic Update

To achieve real-time and consistent unfolding of the colon, it is crucial to update the flattened image F whenever new information becomes available. This need arises as the SLAM pipeline continually refines frames poses, updates frames segment assignment, and copes with real-life issues. To accomplish this, we closely monitor the continuous change in frame poses and their assignment to segments, updating the flattened image through the integration and de-integration of frames. By adopting this strategy, we can rectify errors resulting from registration drift or tracking loss.

Managing Unfolding Updates: When an input frame arrives, we seek to integrate it into the flattened image as quickly as possible, to give the physician instantaneous feedback of the colon coverage. Since previous frames segment-assignments or poses could be updated from [17], we de-integrate and re-integrate all frames if their segment assignment changes. In addition, we sort all frames within each segment in descending order, based on the difference between their previous and updated poses. After sorting, we select and re-integrate the top 10 frames from the list. This allows us to dynamically update the unfolded image to produce a globally-consistent representation of the unfolded colon.

Integration and De-integration: Integration of an RGBD frame f_i is defined as a weighted average of previous mapped samples. For each pixel p in the flattened image, let $F(p)$ denote its color, $W(p)$ the pixel weight, $d_i(p)$ the frame's sample color to be integrated, and $w_i(p)$ the integration weight for a sample of f_i. Each pixel is then updated by:

$$F'(p) = \frac{F(p)W(p) \pm w_i(p)d_i(p)}{W(p) \pm w_i(p)}, W'(p) = W(p) \pm w_i(p), w_i(p) = 1 \quad (2)$$

where the $+$ sign is used for integration and the $-$ for de-integrating a frame. A frame in the flattened image can be updated by de-integrating it from its original pose and/or segment and integrate it with a new pose into its updated segment.

4 Experiments

This section presents the validation of our solution through multiple tests, a comparative study with partially analogous prior works being unfeasible due to the unavailability of essential data and code. The first test, named 'Colon unfolding verification', demonstrates that our 2D flattened visualization is a valid representation of the scanned colon. It also showcases our ability to detect and localize 'holes' in the colon using this visualization. To carry out this test, we used coverage annotations of short colonoscopy clips. Each clip was divided into four quadrants (See Fig. 4), and two experienced physicians tagged each quadrant based on its coverage level ('mostly not covered', 'partially covered', 'mostly covered'). We then used ColNav to estimate the coverage of each quadrant and compared it to the physicians' annotations. The second test focuses on the clinical impacts of using our tool during procedures. We estimate coverage and Polyp Recall (PR) with and without the real-time navigation guidance during the scan to demonstrate the possible benefits of our tool. All datasets used are proprietary.

We conducted all of the tests using a calibrated Olympus CF-H185L/I colonoscope on a 3D printed colon model. The colon model was manufactured by segmenting a CT colon scan from [18] and post-processing it to recover the 3D structure of the colon. The model was fabricated from the final mesh using a 3D printer.

ColNav was run on a high-performance computer equipped with an AMD Ryzen 3960x processor, 128 GB of RAM, and an NVIDIA A6000 GPU. The algorithm ran at a speed of 20 FPS while the live endoscope stream was in its native frequency, enabling real-time usage and guidance during the scans.

The annotations for the first test, and the scans in the second test were performed by physicians who, on average, had 6.5 years of experience and conducted 5000 colonoscopies. We also used the baseline PR experiment (without ColNav) as a standard, and only included physicians with a recall of over 50%.

4.1 Results

Colon Unfolding Verification: Two physicians were asked to annotate the coverage level of 83 short clips captured using our colon model. To assess the annotators agreement, we used weighted Cohen's kappa coefficient [12], due to the ordinal nature of the coverage categories. The resulting weighted kappa score of approximately 60% indicated a "moderate" level of agreement. However, the absolute coverage value given by a single physician might be subjective, making calibration and comparison difficult. Thus, to overcome this issue, we tested for agreement on the relative score of the four quadrants. We used Cohen's kappa to measure the agreement between the physicians on the order of the sorted quadrants based on their coverage score (most covered quadrant is first, least one is last). Cohen's kappa using the relative coverage scores between the two annotators is 84.7% meaning 'almost perfect agreement', which showcases that this comparison method is more suited for the task.

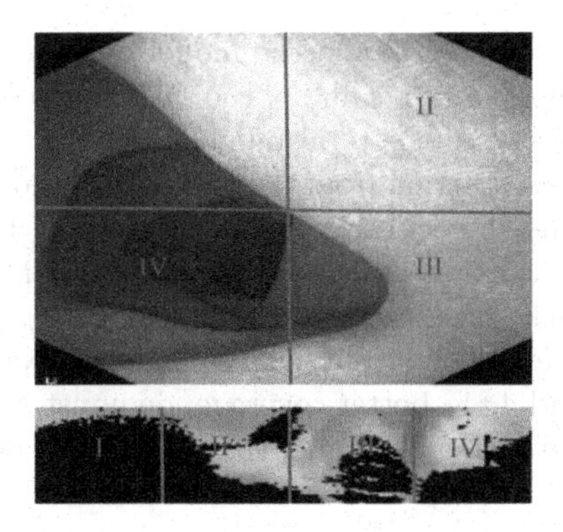

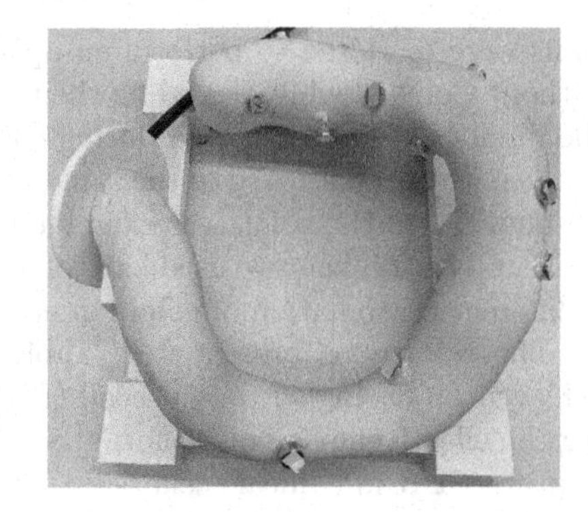

Fig. 4. Left: Representative frame from an annotated clip with the annotated quadrant numbers. Below, ColNav's flattened image of the same clip with the corresponding quadrant numbers. Note that areas that are occluded or aren't visible in the frame are mostly dark in the flattened image. Right: Our complete 3D model with the external magnets that hold in place the small magnetic balls

Table 1. Weighted Cohen's Kappa over the relative coverage scores.

	Cohen's Kappa [%]
Annotators A, B	84.7
Anno. A, **ColNav**	**88.4**
Anno. B, **ColNav**	**85.9**

Table 2. Polyp Recall & Coverage with/out ColNav (*avg.* $\pm$ *std*)

	PR [%]	Cov. [%]
Without ColNav	77.8 $\pm$ 3.9	91.6 $\pm$ 1.5
With **ColNav**	**88.9** $\pm$3.9	**96.4** $\pm$1.0

Based on this approach, we applied ColNav to compute a flattened image for each short clip. Each flattened image was partitioned into four quadrants, and the coverage percentage was calculated for each quadrant. To evaluate our predictions, we mapped the coverage percentages to three categories using a simple threshold: ($cov. <= 60\%$: 'mostly not covered', $60\% < cov. <= 80\%$: 'partially covered', $80\% < cov. <= 100\%$: 'mostly covered').

The results shown in Table 1 present ColNav high levels of agreement with the two physicians, with agreement rates of 88.4% and 85.9% respectively. These results demonstrate that ColNav accurately represents the scanned colon and has high inter-rater reliability with physicians for predicting coverage levels.

Polyp Recall Impact: Real-time estimation of coverage offers the crucial advantage of guiding physicians to potentially missed areas and enhancing the detection of polyps in these regions. To evaluate this capability, we conducted a simulation study by concealing 18 small magnetic balls (diameter = 5 mm) along the full extent of our colon model, thus simulating polyps (see Fig. 4, Fig. 5).

The placement of the polyps was carefully selected so they could be concealed within the colon folds, while remaining unseen unless the endoscope was intentionally maneuvered to inspect those specific areas. Three trained physicians were recruited to perform an optical colonoscopy on the model, with and without ColNav, while recording their coverage and recall, i.e. the ratio between the number of balls detected during each test and the total number of balls.

Scans were conducted in the same manner, starting from the end of the model ('cecum'), and the colon was examined while the endoscope was withdrawn. To prevent location bias, we used balls of multiple colors and assigned each physician a different combination of colors in each test phase (with/without ColNav).

The results, as presented in Table 2, reveal that physicians using ColNav achieved 11.1% higher polyp recall (PR) and 4.8% better coverage, demonstrating the effectiveness of our solution and supporting our belief that ColNav could improve PDR in clinical scenarios.

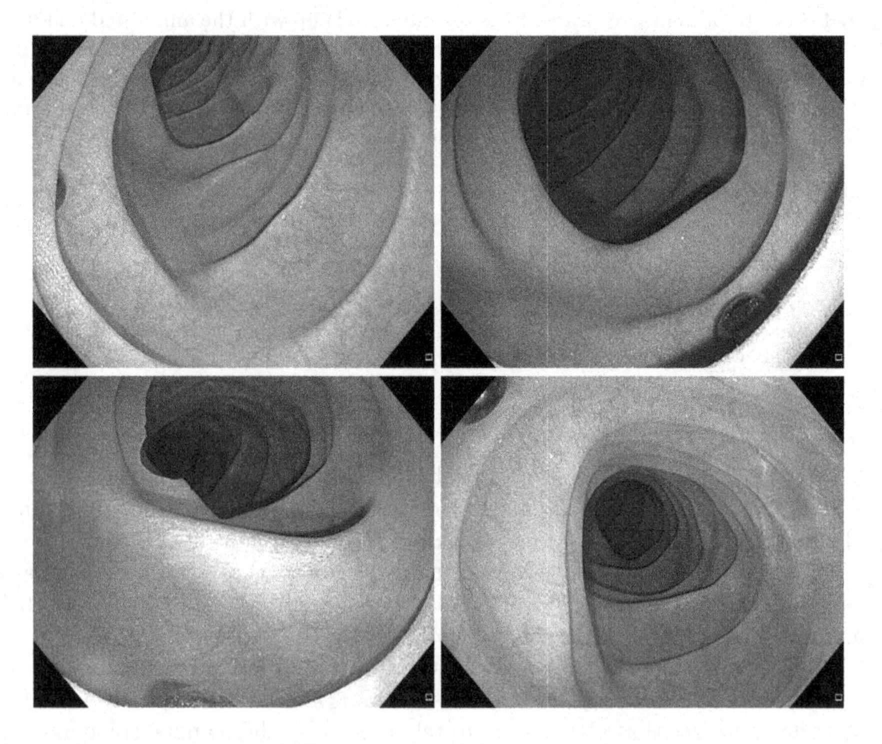

Fig. 5. frames containing hidden magnetic balls in multiple colors, that were used to simulate polyps. Unless the endoscope is intentionally maneuvered to inspect those specific areas the polyps won't be visible

Real Colonoscopy Videos: To evaluate ColNav ability to handle real OC videos, a qualitative analysis was carried out, in which several short clips (a few hundred frames each) of real colonoscopy procedures were selected. ColNav was used on the recorded procedures, to unfold the colon in order to verify that miss-scanned ares can be spotted even in real scans.

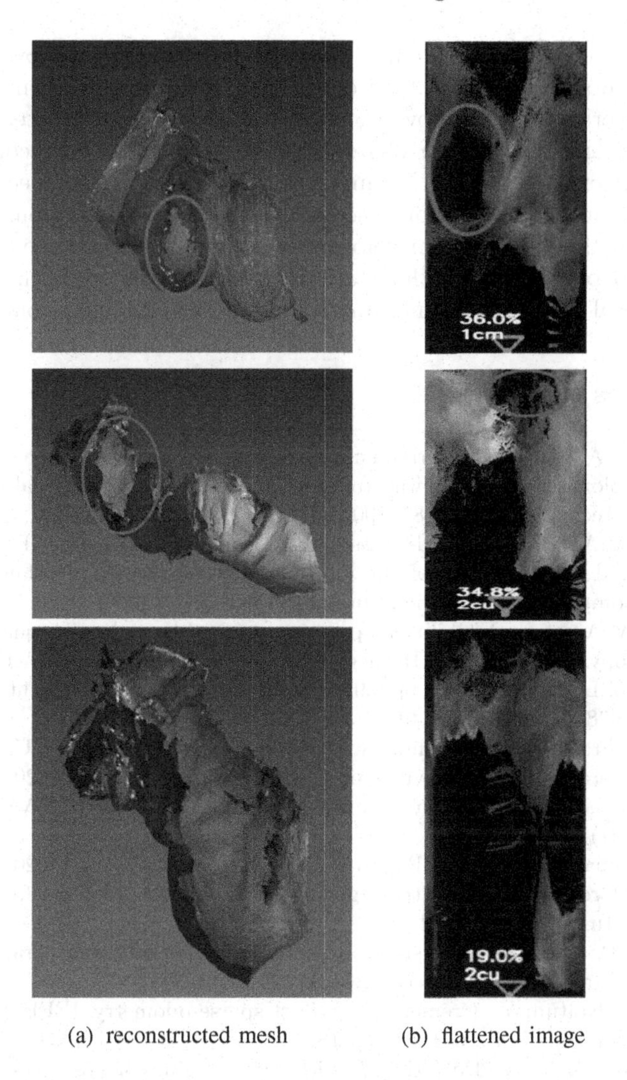

Fig. 6. ColNav on real colonoscopy video segments. (a) 3D reconstructed mesh, (b) flattened image consists of all frames in the sequence. Red ellipses mark corresponding uncovered areas. Best viewed in color. (Color figure online)

Figure 6 shows the 3D reconstruction of the short clip alongside the ColNav flattened colon. From the results, it is evident that the system is able to detect missing areas (circled in red) even in real colonoscopy procedures.

5 Conclusion

We have presented ColNav, the first of its kind real-time colon navigation system, which not only calculates coverage, but also provides augmented guidance to un-surveyed areas without disrupting the procedure. The coverage estimation has been shown to have high correlation with experts. Using the system, physicians were able to improve their coverage and recall in detecting findings within the colon. The system was qualitatively evaluated on recorded real-life procedures. Considering that the employed SLAM module was specifically developed to be robust to optical colonoscopy issues (specular reflections, texture-less areas, motion/fluid blur, minor deformations), our system was able to deal with the real colonoscopy clips without any changes. Further research will focus on improving real-time performance and robustness to extreme colon deformation.

References

1. Armin, M.A., Barnes, N., Grimpen, F., Salvado, O.: Learning colon centreline from optical colonoscopy, a new way to generate a map of the internal colon surface. Healthc. Technol. Lett. **6**, 187–190 (2019)
2. Armin, M.A., Chetty, G., de Visser, H., Dumas, C., Grimpen, F., Salvado, O.: Automated visibility map of the internal colon surface from colonoscopy video. Int. J. Comput. Assist. Radiol. Surg. **11**, 1599–1610 (2016)
3. Armin, M.A., et al.: Visibility map: a new method in evaluation quality of optical colonoscopy. In: Navab, N., Hornegger, J., Wells, W.M., Frangi, A.F. (eds.) MICCAI 2015. LNCS, vol. 9349, pp. 396–404. Springer, Cham (2015). https://doi.org/10.1007/978-3-319-24553-9_49
4. Blau, Y., Freedman, D., Dashinsky, V., Goldenberg, R., Rivlin, E.: Unsupervised 3d shape coverage estimation with applications to colonoscopy. In: 2021 IEEE/CVF International Conference on Computer Vision Workshops (ICCVW), pp. 3364–3374 (2021)
5. International Agency for Research on Cancer: Globocan 2020: Cancer fact sheets-colorectal cancer. https://gco.iarc.fr/today/data/factsheets/cancers/10_8_9-Colorectum-fact-sheet.pdf
6. Dierckx, P.: Algorithms for smoothing data with periodic and parametric splines. Comput. Graph. Image Process. **20**(2), 171–184 (1982)
7. Engel, J., Koltun, V., Cremers, D.: Direct sparse odometry. IEEE Trans. Pattern Anal. Mach. Intell. **40**, 611–625 (2018)
8. Freedman, D., et al.: Detecting deficient coverage in colonoscopies. IEEE Trans. Med. Imaging **39**(11), 3451–3462 (2020)
9. Haker, S., Angenent, S.B., Tannenbaum, A.R., Kikinis, R.: Nondistorting flattening maps and the 3-d visualization of colon ct images. IEEE Trans. Med. Imaging **19**, 665–670 (2000)
10. Liang, Z., Richards, R.: Virtual colonoscopy vs optical colonoscopy. Expert Opin. Med. Diagn. **4**(2), 159–169 (2010). 20473367[pmid]
11. Ma, R., et al.: Rnnslam: reconstructing the 3d colon to visualize missing regions during a colonoscopy. Med. Image Anal. **72**, 102100 (2021)
12. McHugh, M.: Interrater reliability: the kappa statistic. Biochemia medica: asopis Hrvatskoga društva medicinskih biokemčara/HDMB **22**, 276–282 (2012)
13. Mirzaei, H., Panahi, M., Etemad, K., GHanbari-Motlagh, A., Holakouie-Naini, K.A.: Evaluation of pilot colorectal cancer screening programs in Iran. Iran. J. Epidemiol. **12**(3), 21–28 (2016)

14. Mohaghegh, P., Ahmadi, F., Shiravandi, M., Nazari, J.: Participation rate, risk factors, and incidence of colorectal cancer in the screening program among the population covered by the health centers in arak, iran. Int. J. Cancer Manag. **14**(7), e113278 (2021)
15. Moshfeghi, K., Mohammadbeigi, A., Hamedi-Sanani, D., Bahrami, M.: Evaluation the role of nutritional and individual factors in colorectal cancer. Zahedan J. Res. Med. Sci. **13**(4), e93934 (2011)
16. Pore, A., et al.: Colonoscopy navigation using end-to-end deep visuomotor control: a user study. In: 2022 IEEE/RSJ International Conference on Intelligent Robots and Systems (IROS), pp. 9582–9588 (2022)
17. Posner, E., Zholkover, A., Frank, N., Bouhnik, M.: C^3fusion: consistent contrastive colon fusion, towards deep slam in colonoscopy. arXiv:2206.01961 (2022)
18. Smith, K., et al.: Data from ct colonography. Cancer Imaging Arch. (2015)
19. Sudarsky, S., Geiger, B., Chefd'hotel, C., Guendel, L.: Colon unfolding via skeletal subspace deformation. In: Metaxas, D., Axel, L., Fichtinger, G., Székely, G. (eds.) MICCAI 2008. LNCS, vol. 5242, pp. 205–212. Springer, Heidelberg (2008). https://doi.org/10.1007/978-3-540-85990-1_25
20. Vilanova Bartroli, A., Wegenkittl, R., Konig, A., Groller, E.: Nonlinear virtual colon unfolding. In: Proceedings Visualization, VIS 2001, pp. 411–579 (2001). https://doi.org/10.1109/VISUAL.2001.964540
21. Wan, M., Liang, Z., Ke, Q., Hong, L., Bitter, I., Kaufman, A.: Automatic centerline extraction for virtual colonoscopy. IEEE Trans. Med. Imaging **21**(12), 1450–1460 (2002)
22. Wang, G., McFarland, G., Brown, B., Vannier, M.: Gi tract unraveling with curved cross sections. IEEE Trans. Med. Imaging **17**(2), 318–322 (1998)
23. Yao, H., Stidham, R.W., Soroushmehr, R., Gryak, J., Najarian, K.: Automated detection of non-informative frames for colonoscopy through a combination of deep learning and feature extraction. In: 2019 41st Annual International Conference of the IEEE Engineering in Medicine and Biology Society (EMBC), pp. 2402–2406 (2019)
24. Zhou, J., et al.: A novel artificial intelligence system for the assessment of bowel preparation (with video). Gastrointest. Endosc. **91**(2), 428-435.e2 (2020)

Modeling Barrett's Esophagus Progression Using Geometric Variational Autoencoders

Vivien van Veldhuizen[1(✉)], Sharvaree Vadgama[1], Onno de Boer[2], Sybren Meijer[2], and Erik J. Bekkers[1]

[1] University of Amsterdam, Amsterdam, The Netherlands
vivien.van.veldhuizen@hotmail.com
[2] Amsterdam University Medical Centres, Amsterdam, The Netherlands

Abstract. Early detection of Barrett's Esophagus (BE), the only known precursor to Esophageal adenocarcinoma (EAC), is crucial for effectively preventing and treating esophageal cancer. In this work, we investigate the potential of geometric Variational Autoencoders (VAEs) to learn a meaningful latent representation that captures the progression of BE. We show that hyperspherical VAE ($\mathcal{S}$-VAE) and Kendall Shape VAE show improved classification accuracy, reconstruction loss, and generative capacity. Additionally, we present a novel autoencoder architecture that can generate qualitative images without the need for a variational framework while retaining the benefits of an autoencoder, such as improved stability and reconstruction quality.

Keywords: Oncology · Pathology · Variational Autoencoders · Geometric Deep Learning · Equivariance · Representation Learning

1 Introduction

Esophageal adenocarcinoma (EAC) is an aggressive type of cancer with a generally poor prognosis that could benefit from recent advances in machine learning, as it is often diagnosed at a late stage. The only known precursor to EAC, Barrett's Esophagus (BE), progresses through different stages [17] (Fig. 1), providing an opportunity for early detection and prevention. Currently, the detection of dysplasia relies on subjective assessment by pathologists. Advancements in deep learning have introduced the concept of a *digital pathologist* using convolutional neural networks [9,14]. However, while these models have shown promise, they are limited by a high degree of interobserver variability in labeled training data [18]. In this work, we explore the potential of unsupervised learning through various forms of Variational Auto-Encoders (VAEs) [10] in the context of biomarker research. We utilize an unsupervised representation learning approach in order to obtain objective tissue representations and explore to what extent learned representations form a complete description of the tissue by quantifying how

© The Author(s), under exclusive license to Springer Nature Switzerland AG 2023
S. Ali et al. (Eds.): CaPTion 2023, LNCS 14295, pp. 132–142, 2023.
https://doi.org/10.1007/978-3-031-45350-2_11

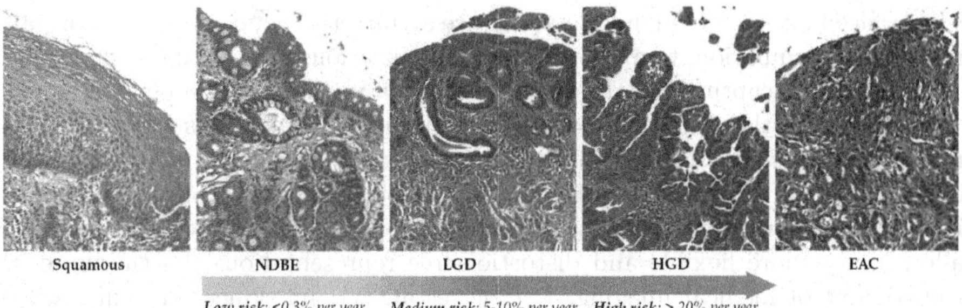

Fig. 1. Different stages of progression: regular squamous epithelium, non-dysplastic BE, low-grade dysplastic BE, high-grade dysplastic BE, and EAC

well an input sample can be reconstructed from the latent representation (**I**), are meaningful in the context of BE by investigating how well classifiers can predict tissue stage, taking only the latent representations as input (**II**), and are interpretable by exploring the generative capabilities of learned models (**III**).

In the context of representation learning, we refer to interpretability as both the capability of generating images from latents (thus providing visual interpretation) and the ability to interpolate between learned representations. That is, in an ideal scenario, the latent space is organized in regions that correspond to different stages of BE, and interpolation would correspond to a smooth transitioning from healthy towards cancerous tissue via NDBE, LGD and HGD. It is known, however, that interpolation using VAEs suffers from latent-space distortion, in which case nonsensical images are generated along the trajectory [4]. This can be avoided through geometric modeling of latent spaces (Sect. 1.1).

In this paper we explore the importance of geometric latent space modeling by comparing hyperspherical VAEs [7] to normal VAEs, and explore a recently proposed *equivariant* variant [16] that allows us to be insensitive to the arbitrary orientation in which tissue is imaged under a microscope [11]. In particular, we address the objectives **I-III** through an extensive empirical study that compares different variants of (V)AEs: variational vs. non-variational; normal Euclidean vs. hyperspherical latent spaces; equivariant vs. equivariant architectures. Additionally, we solve the problem of hyperspherical VAEs being limited to small latent-space sizes, by proposing *a new loss that turns hyperspherical autoencoders into generative models.*

1.1 Related Work

In clinical settings, BE is diagnosed through endoscopic surveillance, where biopsies are taken from the esophagus lining and examined under a microscope. The Vienna criteria [18] are used to classify the severity of dysplasia in BE, which is subdivided into Non-Dysplastic Barrett's Esophagus (NDBE), Low-Grade Dysplasia (LGD), High-Grade Dysplasia (HGD), and an indefinite class for uncertain diagnoses. Pathologists use specific tissue grading features, such as clonality, surface maturation, glandular structure architecture, cytonuclear abnormalities, and inflammation, to make accurate classifications [17]. Such morphological changes can be captured in the latent space of a variational autoencoder.

To mitigate the distortion issue of the original VAE, various VAEs utilizing non-Euclidean manifold have been proposed, such as Riemannian [1,5,12,15], elliptic [2,8,13], or hyperbolic [7]. Notably, Davidson et al. [7] proposes a spherical VAE framework ($\mathcal{S}$-VAE) that operates on a hyperspherical latent space, allowing for more flexible and distortion-free representations. Furthermore, in the context of medical imaging, Lafarge et al. [11] introduced an equivariant VAE model ($SE(2)$-VAE), to tackle the issue of encoding irrelevant information, specifically orientation and translation. The SE(2)-VAE extends the traditional VAE with a group-convolutional neural network [6], enabling the model to be invariant to arbitrary rotations and translations. Building upon these advancements, Vadgama et al. [16] proposed the KS-VAE framework, combining a hyperspherical latent space and an orientation-disentangled group-convolutional network.

2 Method

2.1 VAEs and Hyperspherical VAEs

VAEs [10] are powerful unsupervised learning models based on the assumption that data is generated via $x = D(z) + \epsilon$ with ϵ random noise and D a so-called *decoder*, that decodes the data content from a low-dimensional latent variable. It defines a conditional data distribution $p(x|z)$, the likelihood, which together with a prior distribution $p(z)$ on the latent space defines a distribution on the data space from which one can generate (sample) new data points. One is typically interested in obtaining the compressed latent variable z for a given input x, which can probabilistically be done via the posterior $p(z|x)$. However, the computation of the true posterior is typically intractable and one thus resorts to approximating it with a distribution $q(z|x)$ that is parameterized by an *encoder* neural network E. Via the approximation, one does not directly maximize the (marginal) data-evidence, but instead the Evidence Lower Bound (ELBO) [10]:

$$\mathcal{L}_{\text{ELBO}} = \mathbb{E}_{q(z|x)}[\log p(x|z)] - \text{KL}(q(z|x)||p(z)), \tag{1}$$

which consists of a reconstruction loss (measuring fidelity of reconstructed data) and the KL divergence between the approximate posterior and prior on z.

When the latent space is Euclidean, the approximate posterior $q(z|x)$ and prior $p(z)$ are usually normally distributed, with the parameters of $q(z|x)$ obtained via the encoder network, and those of $p(z)$ set as hyperparameters. For hyperspherical latent spaces we need an equivalent of the normal distribution, which is given by the von Mises-Fisher (vMF) distribution:

$$q(\mathbf{z} \mid \mu, \kappa) = \frac{\kappa^{m/2-1}}{(2\pi)^{m/2}\mathcal{I}_{m/2-1}(\kappa)} \exp\left(\kappa \mu^T \mathbf{z}\right), \tag{2}$$

where the mean μ is a unit vector ($\|\mu\| = 1$), κ is a precision parameter, and $\mathcal{I}_n(\kappa)$ denotes the modified Bessel function of the first kind at order $n = (m/2 - 1)$. For the special case of $\kappa = 0$, the vMF represents a Uniform distribution on the $(m-1)$-dimensional hypersphere $U\left(\mathcal{S}^{m-1}\right)$. The closed form for KL divergence term between a uniform distribution and vMF distribution is derived in [7].

A key element of hyperspherical VAEs is that, due to the compactness of the latent space, it is possible to work with *uniform priors that make sure that the entire latent space is utilized* and every $z \in \mathcal{S}^{m-1}$ corresponds to a sensible data point x. In contrast, in Euclidean VAEs mass in the prior $p(z)$ is typically centered around the origin. Thus, only a fraction of the space is used, resulting in inefficiency and challenges in effectively modeling and separating clusters in the latent space. Hyperspherical models do not suffer from these limitations.

2.2 Generative Hyperspherical Autoencoder Through a New Loss

Hyperspherical VAEs are known to be limited in generative capabilities when the dimensionality of the hypersphere m becomes large, due to instability in sampling from the posterior vMF distributions [7]. We solve this issue by leveraging the fact that, due to the uniform prior, the entire latent space is covered. That is, every $z \in \mathcal{S}^{m-1}$ will equally likely generate a realistic sample of the learned data distribution. We then propose to avoid having to sample during training, by training an autoencoder (usually trained with only the reconstruction loss) with an additional loss that encourages a uniform coverage of data in the latent space which we call the *spread loss.* The spread loss maximizes the distance between encoded data points in a batch via

$$L_{\text{spread}} = \sum_{i,j=1}^{N} -\mathbf{z}_i^T \mathbf{z}_j, \tag{3}$$

where we note that maximizing the true distance $d(\mathbf{z}_i, \mathbf{z}_j) = \arccos(\mathbf{z}_i^T \mathbf{z}_j)$ is equal to minimizing (hence minus sign in (3)) their inner products $\mathbf{z}_i^T \mathbf{z}_j$. An example visualizing the effects of spread loss is shown in Fig. 2.

2.3 Roto-Equivariant VAE and KS-VAE

In addition to exploring different geometric latent spaces, we also investigate the idea of learning orientation-disentangled representations. Classic convolutional neural networks are not equivariant to rotation, causing the same image patches in different orientations to result in different learned representation vectors. Since orientation of scanned biopsies is arbitrary and the intrinsic properties remain unaltered by rotations, we want to learn rotation invariant representations.

We modify (V)AEs to be rotation equivariant based on the method and code of [3]. We note that equivariance means that if the input rotates, the encoded latent transforms in a predictable manner via an action of the rotation group on the latent space. We follow the approach by Vadgama et al. [16] which utilizes an

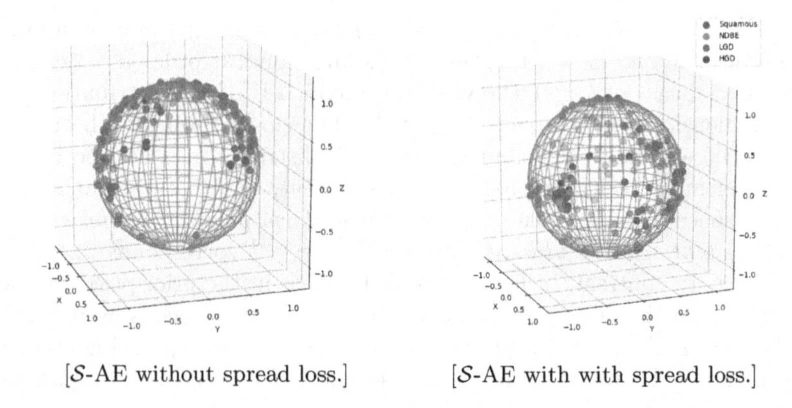

Fig. 2. Visualization of 3-D Latent Space for model $\mathcal{S}$-AE without and with spread loss. The same batch of 200 images was encoded by both models, and different image classes are visualized with different colored points. It can be observed that the points encoded by the model trained with spread loss cover a significantly larger area of the sphere

equivariant encoder to obtain the latent representation z, together with a pose $\mathbf{R} \in SO(2)$ (a rotation matrix), which can be utilized to map z to a canonical pose $z_0 = \rho(\mathbf{R}^{-1})z$, with ρ a representation of the rotation group acting on the latent space S^{m-1}. In their work it is shown that if the hyperspherical latent space is of dimension $(n-1)*2-1$, the latents z can be interpreted and visualized as shapes/visual symbols that consist of n two-dimensional landmarks in a Kendall shape space. The approach is similar to the equivariant VAEs developed by Lafarge et al. [11], except that [16] canonicalizes latents z via a predicted pose $\mathbf{R}$, and that our approach has a hyperspherical instead of Euclidean latent space.

3 Experiments

3.1 Dataset

We train the models on a proprietary dataset retrieved from the Department of Pathology of the Amsterdam University Medical Centers and the LANS-panel (Dutch expert board of esophageal cancer). This dataset consists of digitized and annotated H&E-stained endoscopic biopsies containing different BE progression stages. Additionally, we use the BOLERO dataset, which includes biopsies assessed by a panel of expert BE pathologists [18]. Combining these datasets, we have a total of 934 biopsies from 324 patients. We use the BOLERO dataset as the test data. We also reserve 10% of the training set as a validation dataset.

The biopsies were digitized using a Philips Intellisite Ultrafast scanner and stored as Whole-Slide Images (WSIs), which are highly precise scans of glass slides containing multiple biopsies at various magnification levels. We preprocess the data by dividing the WSIs into smaller patches of size 64×64. See also

Fig. 3. To ensure sufficient context, we choose a magnification level of 5× and only include patches with a threshold of 50% or more relevant tissue (squamous, NDBE, LGD, or HGD classes). Patch labels are computed based on pathologists' annotations in accompanying segmentation files, with the label determined by the dominant class within each patch. To balance the dataset and account for class imbalances, we stratify the dataset by selecting the 8,000 patches for each class, resulting in a balanced dataset of 32,000 patches.

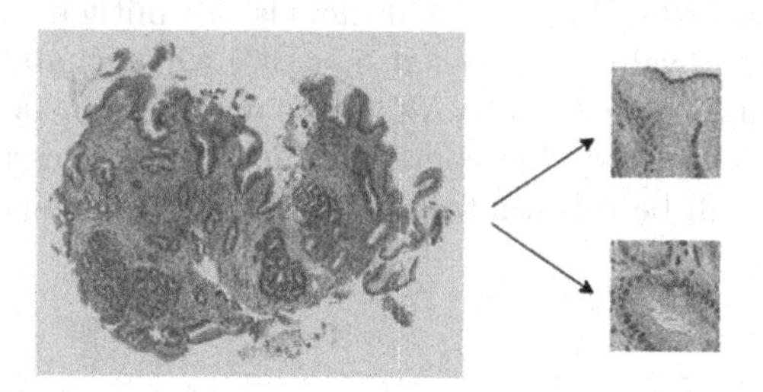

Fig. 3. Example of WSI and extracted image patches

3.2 Experimental Setup

We refer to hyperspherical and normal Euclidean VAEs as $\mathcal{S}$-VAE and *vanilla*-VAE respectively. In our experiments, we investigate representation learning models over three axes: 1) We compare hyperspherical to Euclidean latent spaces, 2) for each model we test both an equivariant (G-CNN) and non-equivariant (standard CNN) version, 3) we compare variational versus non-variational autoencoders. We employ the same architecture for all models, based on the work of Lafarge et al. [11]. The encoder consists of three ConvNeXt blocks followed by max pooling, while the decoder mirrors this structure. The

Table 1. Reconstruction losses on test dataset

	Normal				Spherical			
	Non-Equivariant		Equivariant		Non-Equivariant		Equivariant	
M	VAE	AE	Eq. VAE	Eq. AE	$\mathcal{S}$-VAE	$\mathcal{S}$-AE	Eq. $\mathcal{S}$-VAE	Eq. $\mathcal{S}$-AE
3	1895.56	2013.46	–	–	1930.60	2089.98	–	–
8	1807.06	1769.14	2103.47	2103.35	1707.14	1743.25	2103.66	2103.27
16	1635.89	1621.97	1290.60	1252.64	1563.19	1607.79	1303.07	1313.58
32	1435.60	1428.11	1161.32	1134.30	1389.64	1403.93	1142.24	1135.50
64	1260.85	1273.44	993.94	988.42	1250.45	1258.50	1009.90	992.39
128	1092.88	1113.56	857.40	853.79	1133.33	1104.18	902.82	853.75
256	904.53	935.09	710.22	706.50	1056.68	925.07	826.10	703.88
512	748.42	736.92	562.11	556.61	–	727.06	–	540.13

non-equivariant variational models generate parameters for the relevant posterior distribution, while the equivariant models also predict a pose per sample [16]. We train all models for 500 epochs with a batch size of 128, utilizing the Adam optimizer and MSE loss. We pad and normalize input images, excluding outer edges for equivariant models during reconstruction loss computation.

To fairly compare the models, we vary the latent dimension size and test sizes 3, 8, 16, 32, 64, 128, 256, and 512. As observed in previous research by Davidson et al. [7] and confirmed in our experiments, high dimensions (> 32) pose numerical instability for spherical models. To mitigate this instability, we introduce a minimum value of κ (set at $\kappa = 100$) which enables successful training of spherical autoencoders and VAEs up to a dimension size of 256. However, it is important to note that this approach limits the expressivity of the model, and this trade-off will be taken into account during result analysis.

4 Results

The experiments address three qualities of representation learning with the following questions: (**I**) are the learned representations complete (from a compression perspective); (**II**) are the learned representation semantically meaningful?; (**III**) what are the generative capabilities of each model?

Table 2. Classification Accuracy of Latent Representations on Test Dataset

	Normal						Spherical			
	Non-Equivariant			Equivariant			Non-Equivariant		Equivariant	
M	VAE	AE	CNN	Eq. VAE	Eq. AE	Eq. CNN	$\mathcal{S}$-VAE	$\mathcal{S}$-AE	Eq. $\mathcal{S}$-VAE	Eq. $\mathcal{S}$-AE
3	0.25	0.26	0.46	–	–	–	0.25	0.26	–	–
8	0.33	0.35	*0.48*	0.34	0.17	*0.45*	0.36	0.33	0.23	0.27
16	0.39	0.40	*0.47*	0.39	0.42	*0.52*	0.40	0.34	0.34	0.30
32	0.41	0.40	*0.46*	0.47	0.46	*0.50*	0.41	0.31	**0.49**	0.46
64	0.45	0.40	*0.45*	0.40	0.41	*0.54*	0.39	0.41	0.40	0.43
128	0.42	0.42	*0.47*	0.40	0.40	*0.51*	0.40	0.39	0.40	0.28
256	0.42	0.42	*0.51*	0.38	0.40	*0.50*	0.40	0.41	0.39	0.24
512	0.38	0.36	*0.47*	0.38	0.39	*0.51*	–	0.37	–	0.25

Table 1 addresses (**I**) following the idea that minimal information is lost if the decoder can reconstruct the input from the latent representation z. Here we observe the following: 1) increasing latent dimension size improves reconstruction fidelity; 2) the difference in variational vs non-variational autoencoders is small, but gets more pronounced in the hyperspherical showing that non-variational methods are preferred for compression; 3) equivariant methods have better reconstructions than non-equivariant ones; 4) hyperspherical latent space models outperform Euclidean ones.

Table 2 characterizes the semantic meaning of learned representations (**II**) by testing how well we can train a classifier to categorize a given latent z into each of the classes as given in Fig. 1. As a baseline, we trained a model with the default encoder architecture to directly predict class from the input patch. This should provide an upper bound on classification performance, as this model has access to all available (uncompressed) data to do the classification. The baseline accuracy (upper bound) is 0.51 for non-equivariant and 0.54 for equivariant CNN variants. From Table 2 we make the following observations: 1) Latent dimensions 32 and 64 consistently achieve the highest accuracy across models; 2) hyperspherical VAEs overall give the best performance; 3) the performance of latent space classifiers is close to the upper bound, suggesting that semantic meaning is preserved by the encoders.

To gain insight into the generative capabilities and visual interpretability (**III**) of the learned latent spaces, we sample vectors from random latent locations in all trained models and dimension sizes. Figure 4 showcases one sample per model and dimension size, demonstrating the general differences between the models. In terms of these generated images, a noticeable trend is the decrease in quality for higher dimension sizes across almost all models. Lower dimensions (3, 8, and 16) produce rough, blurry shapes with limited detail. However, in higher dimensions, images become less realistic, losing shapes and introducing colors not present in the original dataset. *The only models capable of generating realistic images consistent with reconstructions in higher dimensions are the spherical VAE and its equivariant counterpart.* Among these models, equivariant $\mathcal{S}$-VAE exhibits slightly more biopsy patch-like structures. *However, even the best models generate images that are too blurry to consider them interpretable.*

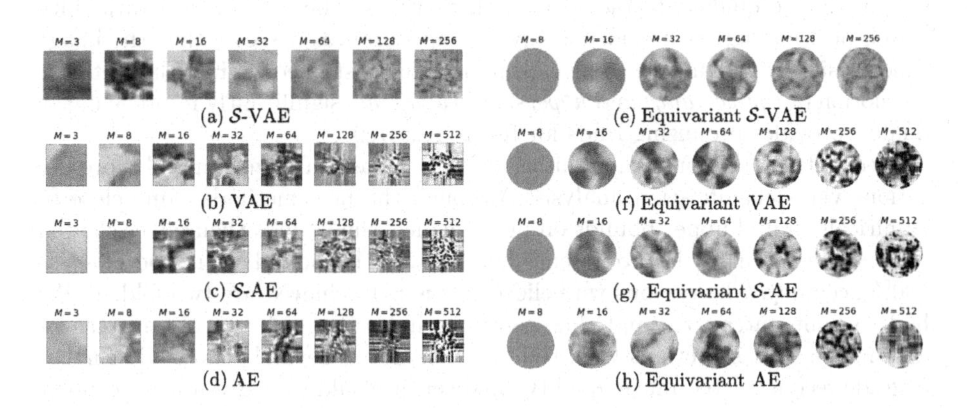

Fig. 4. Randomly generated images from all model types. Each column shows one sample from a model trained with a specific latent dimension size

Finally, to evaluate our novel $\mathcal{S}$-AE model and determine its potential as a generative model, we examine the effects of spread loss on the spherical autoencoder model by evaluating randomly generated images. From these images,

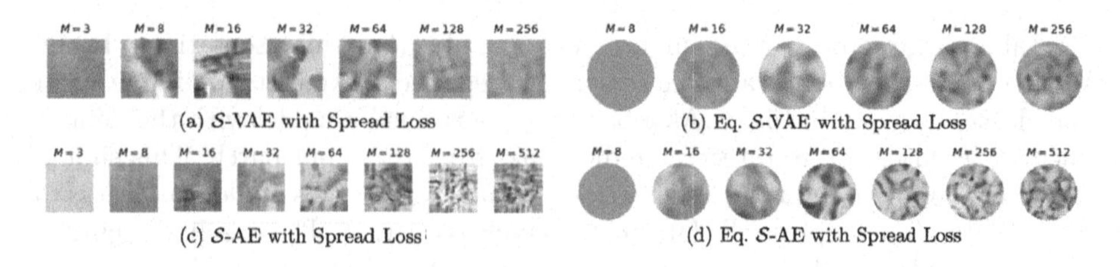

(a) $\mathcal{S}$-VAE with Spread Loss (b) Eq. $\mathcal{S}$-VAE with Spread Loss

(c) $\mathcal{S}$-AE with Spread Loss (d) Eq. $\mathcal{S}$-AE with Spread Loss

Fig. 5. Randomly image samples generated by $\mathcal{S}$-VAE and $\mathcal{S}$-AE with spread Loss, for a range of latent dimension sizes

shown in Fig. 5, it becomes apparent that the introduction of spread loss to the spherical autoencoder substantially improves the quality of generated images. While generated images of autoencoder models previously looked unrealistic, with spread loss they resemble those generated by the variational models.

5 Discussion

Although the experiments provide important insights when it comes to design choices of (V)AEs, which we summarize in the conclusion, we also want to note in what sense the experiments are limited. Firstly, the fidelity of reconstructions and generated samples are not yet at the level one hopes for in a context of interpretability. In comparison, the equivariant VAE of [11], whose neural network architecture we used as a baseline, gave high quality images of single nuclei. However, when scaling up to larger tissue areas, thus including clusters of cells, image quality degrades. We believe this is due to the large variability of cell positionings, their morphology and appearance. It seems that the image space is simply too diverse to be captured with the studied VAEs. The fact that the notion of *equivariance* and *hyperspherical latents* significantly improve image quality provides promising leads for future research.

Secondly, we explored capabilities to learn semantically meaningful representations via a classification analysis. Although the best methods came close to empirically found upper bounds on performance, the bounds themselves showed quite some room for improvement. I.e., ideally, the bound would be close to 100% accuracy. The reasons we believe this is not achieved are two-fold. 1) We had to limit patch-size (and thus context window) in order to obtain reasonable image reconstructions with the (V)AEs. Going beyond this would further degrade reconstructed image quality, however, it would have given more context for the baseline classifier. 2) The labeling of tissue patches is a highly variable and subjective manual task. This is precisely the motivation for why we are investigating unsupervised learning methods. Nevertheless, the experiments show that neural networks can pick up on consistent semantic cues in an unsupervised manner.

6 Conclusion

In this study, we explored the application of several variants of VAE to learn tissue representations in an unsupervised manner, with the intent to develop tools that contribute to an objective understanding of the progression of Barrett's esophagus. Our contributions are threefold: 1) the experimental analysis of (V)AE variants showed the importance of *equivariance* and *hyperspherical latent space* modeling; 2) it showed the potential (latent representations can be semantically meaningful) and limitations (image generations show room for improvement) of generative unsupervised representation learning; and 3) we showed that one can train generative autoencoders in a non-variational setting without compromising on performance. Our novel spread loss allowed to train generative autoencoders without having to rely on a sampling, thereby circumventing the problem of limited latent space dimension of hyperspherical VAEs. Our study showed the stability of generative models with hyperspherical latent spaces and establishes a strong basis for further representation analysis via e.g., cluster analysis or interpolation experiments. We presented first steps towards a quantitative understanding of the latent space of esophageal tissue and how it could be organized along the axis of progression from healthy to cancerous tissue.

References

1. Arvanitidis, G., Hansen, L.K., Hauberg, S.: Latent space oddity: on the curvature of deep generative models. arXiv preprint arXiv:1710.11379 (2017)
2. Bachmann, G., Bécigneul, G., Ganea, O.: Constant curvature graph convolutional networks. In: International Conference on Machine Learning, pp. 486–496. PMLR (2020)
3. Bekkers, E.J., Lafarge, M.W., Veta, M., Eppenhof, K.A.J., Pluim, J.P.W., Duits, R.: Roto-translation covariant convolutional networks for medical image analysis. In: Frangi, A.F., Schnabel, J.A., Davatzikos, C., Alberola-López, C., Fichtinger, G. (eds.) MICCAI 2018. LNCS, vol. 11070, pp. 440–448. Springer, Cham (2018). https://doi.org/10.1007/978-3-030-00928-1_50
4. Chadebec, C., Mantoux, C., Allassonnière, S.: Geometry-aware hamiltonian variational auto-encoder (2020)
5. Chen, N., Klushyn, A., Kurle, R., Jiang, X., Bayer, J., Smagt, P.: Metrics for deep generative models. In: International Conference on Artificial Intelligence and Statistics, pp. 1540–1550. PMLR (2018)
6. Cohen, T., Welling, M.: Group equivariant convolutional networks. In: International Conference on Machine Learning, pp. 2990–2999. PMLR (2016)
7. Davidson, T.R., Falorsi, L., De Cao, N., Kipf, T., Tomczak, J.M.: Hyperspherical variational auto-encoders. arXiv preprint arXiv:1804.00891 (2018)
8. Gu, A., Sala, F., Gunel, B., Ré, C.: Learning mixed-curvature representations in product spaces. In: International Conference on Learning Representations (2018)
9. Hussein, M., et al.: A new artificial intelligence system successfully detects and localises early neoplasia in barrett's esophagus by using convolutional neural networks. United Eur. Gastroenterol. J. **10**(6), 528–537 (2022)
10. Kingma, D.P., Welling, M.: Auto-encoding variational bayes. arXiv preprint arXiv:1312.6114 (2013)

11. Lafarge, M.W., Pluim, J.P., Veta, M.: Orientation-disentangled unsupervised representation learning for computational pathology. arXiv preprint arXiv:2008.11673 (2020)
12. Shao, H., Kumar, A., Thomas Fletcher, P.: The riemannian geometry of deep generative models. In: Proceedings of the IEEE Conference on Computer Vision and Pattern Recognition Workshops, pp. 315–323 (2018)
13. Skopek, O., Ganea, O.E., Bécigneul, G.: Mixed-curvature variational autoencoders. arXiv preprint arXiv:1911.08411 (2019)
14. de Souza Jr, L.A., et al.: A survey on barrett's esophagus analysis using machine learning. Comput. Biol. Med. **96**, 203–213 (2018)
15. Tosi, A., Hauberg, S., Vellido, A., Lawrence, N.D.: Metrics for probabilistic geometries. arXiv preprint arXiv:1411.7432 (2014)
16. Vadgama, S., Tomczak, J.M., Bekkers, E.J.: Kendall shape-vae: learning shapes in a generative framework. In: NeurIPS 2022 Workshop on Symmetry and Geometry in Neural Representations (2022)
17. Van der Wel, M., Jansen, M., Vieth, M., Meijer, S.: What makes an expert barret's histopathologist?, vol. 908, pp. 137–159 (2016)
18. van der Wel, M.J., Coleman, H.G., Bergman, J.J., Jansen, M., Meijer, S.L.: Histopathologist features predictive of diagnostic concordance at expert level among a large international sample of pathologists diagnosing barret's dysplasia using digital pathology. Gut **69**(5), 811–822 (2020)

Author Index

© The Editor(s) (if applicable) and The Author(s), under exclusive license
to Springer Nature Switzerland AG 2023

S. Ali et al. (Eds.): CaPTion 2023, LNCS 14295, pp. 143–144, 2023.
https://doi.org/10.1007/978-3-031-45350-2

Printed in the United States
by Baker & Taylor Publisher Services

Printed in the United States
by Baker & Taylor Publisher Services